GW00634480

FAMOUS FLIGHTS

FAMOUS FLIGHTS

John Frayn Turner

ARTHUR BARKER LIMITED LONDON
A subsidiary of Weidenfeld (Publishers) Limited

Published in Great Britain by
Arthur Barker Limited
91 Clapham High Street, London SW4

ISBN 0 213 16688 7

Printed in Great Britain by
Bristol Typesetting Co Ltd,
Barton Manor, St Philips, Bristol

CONTENTS

TO DOUGLAS BADER

1

THE WRIGHT BROTHERS

the birth of powered flight

THE seventeenth of December 1903 was the day that man first flew an aeroplane – when the Wright brothers made the very earliest sustained, controlled flights in a powered plane at Kitty Hawk, Carolina, USA.

This brief but historic event, a week before Christmas, was the culmination of years of thought and effort not only by the Wrights but by other aeronautical pioneers.

Wilbur and Orville Wright of Dayton, Ohio, began to take a passionate interest in the problems and possibilities of flight from the time of the death of the great German gliding pioneer Lilienthal in 1896.

Starting from the findings of Lilienthal they determined to develop further the art of gliding and eventually to provide power to drive and control flight and make it more or less independent of the elements. But before they could consider powered flight, they had to get as much practical knowledge as possible of the behaviour of gliders.

They kept their bicycle business going, but spent every spare second building a glider. They found the perfect place to test it at Kitty Hawk, Carolina, which gave them the climatic conditions they sought: a steady prevailing wind blowing at around twenty miles an hour. This small settlement stood on a long bleak sand-bar that bridged the waters of Albemarle Sound and the Atlantic Ocean.

It was there in September and October 1900 that the Wrights started the series of trials which would have their climax three years later. The trio of sandhills they chose for the work were about thirty, eighty and one hundred feet high. That first glider was a biplane with an area of 165 square feet, and an eighteen-foot wing span designed and built from the findings of Lilienthal and others, and intended to fly with a man in a wind of anything over 20 miles an hour.

The Wrights tried it out first as a kind of man-carrying kite-cum-glider. Either Wilbur or Orville would lie flat on the middle of the lower wing to minimize head resistance, and then try to fly from one of the sandhills selected. But although the wind blew at 25–30 miles an hour they found that no matter how they tried, they could not get the upward thrust anticipated, and that the pilot could not control the glider properly.

So they started to modify their ideas, substituting weights for the pilot and flying the glider as a kite. This enabled them to measure the forces operating on the glider in various conditions. They were eventually compelled to the conclusion that this design would never soar but only glide short distances downhill. When they did resume gliding tests again, they took every precaution possible against accidents, although both they and the glider did have one or two close calls in their free glides a couple of yards off the sand.

But by the last week of October 1900, they realized that they had learned all they could from glider number one. Anyway they were getting a little worried about the business, and the weather showed signs of deterioration.

Throughout that winter at Dayton, they busily built glider number two, which was bigger than its predecessor. They increased the wing span to twenty-two feet; the weight was nearly double, at just under 100 pounds; and it had an overall area of 308 square feet. This glider had the distinction of being the biggest ever made to date. Among other changes, the Wrights decided to try a greater curvature of the wings.

In late July 1901, they carted the new version to Kitty Hawk on what was the equivalent of their summer holidays. The initial test results disappointed them, for it seemed that they had curved the wings too much and produced a loss of control. Each successive test was followed by trial and error modification, but gradually they were coming to understand the concept of control by balancing planes and rudders.

Their glides in August did in fact gradually grow longer in distance and time, but they still felt dissatisfied and made the long trek home to Dayton pretty depressed.

They next turned their attention to the most vital question of all – wing design. They started to devise and carry out small-scale tests with model wings in a wind tunnel they rigged up at home. The tunnel measured about five feet long and sixteen inches square, and by injecting a flow of air from a fan they could observe the reactions of it on miniature metal wings which they developed literally daily.

In this tiny tunnel they could simulate wind speeds of 25–30 miles an hour and see precisely what effect these produced on as many as two hundred different designs they tried. For hour after hour that winter the Wrights made notes and readings about the various shapes of model wings. And from all the thousands of notes emerged the design of glider number three.

An impartial observer would have been able to detect signs of real advance, but the two brothers were too close to it all and too involved to be able to see it in proper perspective. They only knew they had to fly – somehow.

The wing span went up from twenty-two feet to thirty-two. This extra width helped to give it greater hoisting power. Other innovations included a double fixed vertical tail and a front elevator. The area of the new glider was about the same as the previous one, at 305 square feet, and it weighed some 116 pounds.

The Wright brothers felt more optimistic this summer and first tried out the glider at Kitty Hawk in September 1902. Kitty Hawk had become their summer residence, but the weeks they spent there were far from a holiday. From dawn to sunset they slogged away, flying the glider again and again.

A few days after their arrival Orville Wright was the pilot during one glide when the right wing started to rise. Trying to control it, however, he made it worse and the wing went on rising, throwing the glider more and more out of balance. The whole glider was now tilted up at a nasty angle.

While Orville grappled with the controls, his brother and a

few onlookers below suddenly saw the glider stall and then float backwards and downwards. This was the sort of situation the brothers had struggled to avoid, yet here it was, and those on the ground could only call out vainly. Orville and the glider hit the ground in a spray of sand. The others ran to the wreckage, but luckily Orville was not hurt at all.

By the following week they had repaired the glider, or rather reassembled it, and at the same time they set about remedying what had gone wrong by changing the double tail into one movable fin. They now reckoned that they could control the glider reasonably in three dimensions – and proceeded to prove it.

Wilbur Wright said:

> 'We made nearly 700 glides in the two or three weeks following. When properly applied, the means of control proved to possess a mastery over the forces tending to disturb the equilibrium. We flew it in calms and we flew it in winds as high as 35 miles an hour. We steered it to right and to left, and performed all the evolutions necessary for flight. The machine seems to have reached a higher state of development than the operators; as yet we consider ourselves little more than novices in its management.'

Before they left Kitty Hawk that autumn, the brothers had made no fewer than a thousand actual glides averaging about fifteen seconds each. In one hectic week they clocked up over 375 individual glides, one exceeding 600 feet and lasting nearly half a minute. For those days, that represented an eternity. Most important of all, however, they could now control their flights.

Their next planned step was to add an engine to produce a powered plane – and make the first powered flight of all time. At once a fresh set of difficulties appeared, for as the brothers said subsequently:

> 'What at first seemed a simple problem became more and more complex the more we studied it. With the machine

moving forward, the air flying backward, the propellers turning sideways, and nothing standing still, it seemed impossible to find a starting point from which to trace the various simultaneous reactions.

'Contemplation of it was confusing. After long arguments we often found ourselves in the ludicrous position of each having been converted to the other's side, with no more agreement than when the discussion began.'

The brothers had not only to design and build an aeroplane, but the engine as well, for when they looked around for a suitable petrol engine on the market, none existed. And at the same time they had to experiment with propellers, for next to nothing was known about the precise behaviour of these strange new devices.

First they designed the aeroplane. They based their design on the third version of the glider, extending its linen-covered wing area still further to 510 square feet. It was a wooden-frame biplane spanning just over forty feet and measuring twenty-one feet long, with a twin elevator in front and a twin rudder at the rear.

The Wrights then set about making the components that would produce powered flight – the engine and propellers. They built a four-cylinder engine developing some twelve horse-power at 1,200 revolutions a minute which they mounted on its side on the lower wing in a position calculated to counter-effect the pilot's weight.

They christened the aeroplane the Flyer and took it to Kitty Hawk well crated towards the end of September 1903, together with glider number three. They were full of excitement at the prospects ahead of them, but they had to be superhumanly patient. Stormy weather all through that autumn made it out of the question to test the Flyer, but they went on with their gliding, getting in all the practice they could before the first attempt to fly a powered aircraft – and sustain and control it.

The wild weather abated at last, and the Wrights made ready on 14 December to test the Flyer. The two undercarriage skids of the machine were placed on a trolley, which in turn ran on a monorail. A wire would hold back the Flyer while the engine revved up; the wire would be slipped; the Flyer would thrust forward and, with any luck, be airborne. That was the theory.

Wilbur and Orville tossed up to see who should have that first flight. Wilbur called correctly, but was less lucky with the actual attempt. The Flyer shuddered down the rail, all right, but went up in the air too steeply, stalled, and then crashed. The repairs took a couple of days.

Despite this further setback the Wrights remained so completely confident they were about to make history that they invited the few locals of Kitty Hawk to come and witness the event. Five turned up. Three men from the local life-saving station a mile or so away, a lumber buyer and a sixteen-year-old boy. This was the scene on those desolate dunes, as described by Orville Wright:

'During the night of December 16, 1903, a strong, cold wind blew from the north. When we arose on the morning of the 17th, the puddles of water, which had been standing about camp since the recent rains, were covered with ice. The wind had a velocity of 22 to 27 miles an hour. We thought it would die down before long, but when ten o'clock arrived and the wind was as brisk as ever, we decided to get the machine out.

'Wilbur ran at the side, holding the wing to balance it on the track. The machine, facing a 27-mile wind, started very slowly. The course of the flight up and down was exceedingly erratic, partly due to the irregularity of the air and partly to lack of experience in handling this machine.

'The control of the elevator was difficult on account of its being balanced too near the centre. This gave it a tendency to turn itself when started, so that it turned too far on one side and then too far on the other. As a result the machine would rise suddenly to about 10 feet, and then as suddenly dart for the ground. A sudden dart when a little over 100 feet from the end of the track or a little over 120 feet from the point at which it rose into the air, ended the flight.

'This flight lasted only 12 seconds, but it was nevertheless the first in the history of the world in which a machine carrying a man had raised itself by its own power into the air in full flight, had sailed forward without reduction of speed, and had finally landed at a point as high as that from which it had started.'

One of the men, John T. Daniels, had clicked a camera aimed at the end of the monorail runway and recorded for all time the Flyer in flight, with Orville Wright lying prone on the lower wing. In the photograph, beside the starboard wing, trots Wilbur Wright in a peak-cap, just having let go of the wing and now surely willing the Flyer forward with all his soul. And there too is the Flyer itself, a maze of struts and wires and wings, with no wheels, its skids some three feet off the ground. And it is flying.

It was just after 10.30 am on that memorable morning. Then Wilbur took over the controls, while Orville watched and guided. Again the Flyer sailed forward at some 30 miles an hour for eleven seconds. Orville did a third trip for fifteen seconds, and then came the fourth and longest flight of the day.

The time was noon. Wilbur shot forward and upwards. The Flyer steadied and flew on at a good 30 miles an hour, forcing itself forward in the icy Atlantic wind.

A quarter of a minute passed, then half a minute, three-quarters. Then Wilbur made rather too sharp an adjustment after negotiating a slight sand hillock. The Flyer dipped, dived, and struck the sand. But by then it had travelled 852 feet from its starting place and remained airborne for fifty-nine seconds.

A few minutes later, a gust of wind caught the Flyer, overturned it, and did damage to various vital parts of it. But although the machine never flew again, the air age had arrived.

The fantastic fact, though, was that none of the press were there to see it – and no one really took much notice of it at the time.

It was not really until five years later in France that the world realized the significance of the Wrights' achievements, when throughout that year, Wilbur Wright astounded Europe by a succession of sensational flights, culminating in his record at Le Mans on 31 December 1908, of 2 hours 20 minutes 23.2 seconds in the air. It was no accident then that the next famous flight, in the following year, was by a Frenchman, Monsieur Blériot.

2

BLÉRIOT

first man to fly the Channel

A THOUSAND pounds to the first person to fly the English Channel. That was the offer Lord Northcliffe made in the *Daily Mail* on 5 October 1908. Despite the achievements of the Wright brothers, only a handful of people really recognized the potentials of the aeroplane – either for peace or war. Northcliffe was one of these.

This offer of a prize for the Channel flight fired the imagination of both the public as well as that of the select band of pioneer aviators in Europe at the time. None of these pioneers wasted any time in trying to win the award – and the prestige deriving from the first successful flight over such a famous stretch of water as the English Channel.

Blériot, who had already covered considerable distances in his No XI machine, was determined to win this honour for France. But two more airmen appeared on the scene and Blériot realized he would have to hurry if he were to be the victor.

The odds were against him from the outset, for he had suffered quite a bad crash only a matter of days before he finally attempted the Channel project. Blériot's usual method of extricating himself from fatal or serious injuries was to clamber out on to one of the wings just before crashing, but he had not managed to make his customary exit from his seat on this occasion. One of the inevitable faults had occurred when he was in the air: the fuel pipe snapped and the resulting flow of petrol had flared up frighteningly, forcing him to stay in his seat and bring the plane down as best he could. The blazing fuel ignited the flimsy fuselage and the aeroplane hit the ground in a shambles of smoke and fire. The heat had burned one of his feet badly, but he managed to pitch out of the plane as his helpers hurried up to it.

At the time of the Channel flight he still had the foot bandaged and walked with a limp. Nevertheless he protested he would be all right to fly.

Meanwhile one of his rivals beat him to the first shot at the goal on 19 July. Since the Channel separates England from France, it seemed appropriate that the man making the opening move in the exciting experiment was half English and half French.

His name was Hubert Latham, and he had an English father and French mother. Latham had the use of an elegant Antoinette IV monoplane, distinguished by its slender lines. For those early times, the plane looked unusually graceful, as its name might suggest.

Latham got his crated monoplane to a place called Sangatte, not far from Calais. He assembled it in one of a small group of sheds originally constructed in connection with the scheme for building a Channel Tunnel. Yes, they were talking about it as long ago as that and even earlier! But now the idea had been temporarily shelved and the sheds stood neglected. It was significant, therefore, that he should have chosen a shed associated with this submarine means of bridging the Channel when he was about to try and demonstrate a more effective and expeditious way – above the water.

Latham knew he had to make his effort promptly to stand a chance of winning. So soon after daybreak on 19 July 1909, he helped to get the Antoinette out of the shed and in a matter of minutes he was ready for the flight. He carried a camera to record any details he could of the flight, and he glanced out over the cliffs to see the French destroyer which was standing by some way out in case of accidents. And with these primitive aeroplanes, there usually were.

Latham rose up over Sangatte, circled, and headed out to the calm sea. He set his compass in the direction of Dover and sat back and hoped for the best. The hard work had gone on before, while the aeroplane was being assembled and tested.

He had flown it only a day or so previously and it had seemed to be working well.

But before seven o'clock that morning, when Latham had only flown a very few miles, he began to have sparking plug trouble. The engine started spluttering ominously, missing, and eventually failed completely. As the propeller slowly stopped turning, Latham realized with a jolt that he had to set the machine down somehow on the water.

He glided gently down over the smooth surface at an angle as small as he could possibly manage. Fortunately he was a skilled pilot; the sea scarcely rippled; a French destroyer was steaming towards the area, churning up the sea in its haste.

So six miles out from the Calais coastline, Hubert Latham juggled with his controls to keep the plane as level as possible – and came down to water-level. It would have been a marvellous landing if the ground had not been liquid. At first, the sleek Antoinette fuselage floated on the surface and the wings hardly got wet, but soon the aeroplane did start to submerge slowly, so Latham calmly climbed out on to a wing, where he waited for rescue, which soon came. So expertly had Latham set the plane down on the sea that he remained completely dry the whole time, but the aeroplane was virtually a write-off.

Latham sent an urgent message to Paris for a replacement and was glad to see one arrive at Sangatte only three days after his abortive attempt. It was now 22 July and the race was really on. Latham and his team started to assemble and test the second model at a frantic pace, for he knew that Blériot had by then appeared and that this rival would be likely to prove highly dangerous to his hopes.

The second of Blériot's rivals was some miles off at Wissant, near Boulogne. He bore the imposing title of Comte de Lambert, but despite this and a Wright biplane, the other two did not really see him as a serious contender.

Blériot had heard, of course, of Latham's fall into the water, but this did not deter him at all. He had had far worse spills himself. And hadn't he actually flown his No XI machine a distance equal to the Channel crossing quite recently? The

Blériot No XI was a monoplane, like Latham's, and powered by a 25-horsepower Anzani engine. Now it had been uncrated, assembled, and awaited its moment amid the tufty sand-dune coastline at Baraques, just outside Calais.

After Blériot's wife had bade him farewell, she was taken aboard the French destroyer on rescue duty. There would be no problem in recovering an airman up to about ten miles from the French coast, but if an aeroplane should come down after flying about halfway across the Channel, the destroyer would have been outpaced. Blériot reckoned to fly at a speed around 50 miles an hour, far faster than any ship could manage at that time. So there were very real hazards in this venture. Twenty-odd miles of sea can seem an ocean to a pilot in trouble.

The two chief opponents had been right about the Comte. He was not yet ready to try his luck.

By Saturday, 24 July, both Latham and Blériot were ready for the flight. Then the weather intervened quite dramatically. All day on 24 July the wind whined up the Channel from the south-west, so that neither of them could possibly hope to succeed in their little light aeroplanes. The weather did not look likely to change much as dusk fell on that day, but Blériot felt that the prospects could improve during the night, so he made arrangements to arise well before dawn and travel by motor car the short way to Baraques to be ready in case any attempt were possible on Sunday, 25 July.

As the night died, so did the wind.

The rival machines were housed only a mile or so from each other, and the first that the Latham team knew of any activity by Blériot was when Latham's friends actually glimpsed No XI being wheeled out on the sand flats below the cliffs. Through the dawn haze, they saw four men steadily pushing the little monoplane along on its spoked wheels. The curved propeller quivered in the faint breeze. And the pilot himself stood in the cockpit, his helmet flap dangling under his chin, superintending the movement of the machine towards its starting point. Then he sat down between and behind the two broad wings, and made last-minute checks to his controls, clearly anxious to be up and away.

Before the final formalities, Blériot turned to one of his friends and asked in his usual ultra-casual way: 'Incidentally, where *is* Dover?'

'Over there,' came the response, with an equally casual wave in the general direction of the English coast. Blériot did not carry a compass. It was all a bit haphazard! With that Blériot was ready to take off.

They swung the large propeller manually and the motor started. An engine of the kind in this aeroplane ran without becoming badly hot in the realms of twenty minutes. But Blériot would need twice that time to cross the Channel. That was a measure of the risk involved, though there were others as well.

The official take-off time read: 4.41 am, 25 July 1909. Most of Europe was still fast asleep, unaware that history was about to be made.

Before the Antoinette's team could even call Latham to his machine, and still scarcely believing that Blériot was attempting anything more than a trial, they saw their rival's aeroplane signalled away, ascend over the sands, and set its nose out to sea on the course so casually indicated.

'He's heading for Dover,' one of them cried out in panic. But by then they knew it was already too late to do anything, so they merely watched mesmerized as the Blériot XI veered round and vanished into the morning mist over the Channel.

The grey light showed through the open framework of the fuselage, which ended in the upturned tail at the rear. And the same light pierced the three landing wheels which had stopped spinning by now and would not be able to turn again till they made contact with the ground of another country – England.

Blériot vaguely made out the smoke of the destroyer *Escopette* ahead and below, but then the little monoplane sailed steadily up to, and past, the warship. For a moment or two, Blériot continued to turn round to see if she were still visible, but after ten minutes' flying he realized she could help him no more: he was on his own. From now on if the aeroplane failed and fell he would probably drown.

By this time, the French coast had also faded far behind him and he could not see anything ahead of him. At his low altitude visibility was limited to a few miles anyway. So there he was, surrounded by morning mist; below him, the sea; and somewhere ahead, England – he hoped.

He did not feel too happy knowing he was the first man to be suspended in space in this unhealthy position miles out to sea. With each passing moment, the destroyer fell further astern.

'I was amazed,' he said. 'I could see nothing at all. For ten minutes I was lost. It was a strange position to be alone, unguided, in the air over the middle of the Channel.'

In fact he flew the aeroplane along the approximate line indicated to him at the start by the destroyer. Not exactly advanced navigation but at least an improvement on the original briefing!

Twenty minutes had passed and he was barely half way across. That was the average time the engine was expected to run without showing signs of overheating. Could it hold out?

Then the engine began bumping. He sucked in his breath through his great moustache. He knew the signs. The scorching engine; the audible reaction; the effect on the aeroplane's performance. But luckily he ran right into some gentle drizzle which cooled the small engine sufficiently for it to go on chugging and churning away.

Blériot flew on.

Thirty minutes gone.

Suddenly he saw the rippling silhouette of the clifftops of Dover rising into shape several miles ahead of him. England! He was within sight of success. No one had ever been more delighted to glimpse those famous cliffs. Surely he couldn't fail now? That half-hour had seemed endless – but the final ten minutes were worse.

As he approached the coast he realized that he had been

blown east by the breeze and was heading straight for St Margaret's Bay. He swung the aeroplane round to the south-west towards Dover, where he had arranged for a landing-place to be marked for him by a French journalist. The method of identification was for a French tricolour to be exposed flat on the ground, clearly visible from the air.

But the wind was fiercer now, and in this area where it blew off the sea and over the cliffs, it created currents that made the little aeroplane increasingly awkward to handle. Blériot zoomed up over the actual clifftop, not missing it by much. He could not really control the aeroplane properly, so decided to come down quickly in the first possible place. His leather-helmeted head peered out over the side of the aeroplane and he settled for a stretch of green grass not far from the cliffs where Dover Castle stood proud.

The aeroplane had done its duty well, and Blériot brought it down thankfully. The landing-strip turned out to be a steep slope just across a meadow from the castle, and the machine came down with a considerable crack. The front wheels spreadeagled and smashed, and with them went the propeller. The heavy landing gave Blériot a sharp shock and hurt him a little, but not enough to stop him vaulting out of the damaged machine and taking stock.

No one was in sight!

Several minutes later a policeman came up breathlessly, followed by the French journalist, till gradually a knot of onlookers had arrived to see this strange French flier on English soil. Where did he come from? There was a sailor, a straw-hatted man, several cloth-capped workers, and as it was England – a customs officer! And in the midst of them stood the triumphant, boyish Blériot. He'd done it. Thirty-one miles in forty minutes.

Lord Northcliffe, in a moment of vision, said at once: 'Britain is no longer an island.' How right he was.

And if men could cross the English Channel by air, why not the Atlantic Ocean? Ten years later they did.

3

HAWKER AND GRIEVE

attempt on the Atlantic

IT was just after three o'clock on the afternoon of Sunday, 18 May 1919. Two Englishmen clambered into the cockpits of their Sopwith biplane at Mount Pearl airfield, St Johns, Newfoundland, hoping to be the first to fly the Atlantic non-stop. Aptly enough their speck of a plane was christened *Atlantic.*

The pilot was Harry Hawker and his navigator Commander Mackenzie Grieve. Hawker had flown throughout the four years of the Great War. For weeks they had been waiting for the weather to improve. Now it seemed perfect.

Hawker had made several last-minute changes in his aeroplane, including replacing the original four-bladed propeller by a more conventional two-bladed one, and planning to take the highly unconventional step of jettisoning the undercarriage on take-off to give the machine maximum lift. To lessen the risk of a crash on landing Hawker fitted small steel skids which he hoped would enable the machine to glide along the ground to a stop.

But the landing was far in the future. Now it was 3.15 pm as the two men shook hands and waved the usual farewells to the crowd on the airfield. The sky was blue and the wind blew from the north-west as the little aeroplane jolted off on its long take-off run. It was carrying a huge load of fuel, and the spectators wondered whether it could ever take to the air.

But it did, though to get it up at all, Hawker had to taxi it diagonally across the field for extra length. Finally he managed to coax it into a laborious climb clear of the fence at the edge of the airfield – and they were away.

The sun glinted on the biplane, with its tail waggling slightly as if in pleasure at being airborne. And on that tail was the name Sopwith Aviation Company, Kingston on Thames.

Hawker and Grieve were really on their way, the first men ever to try and fly the Atlantic in a powered aeroplane.

The aircraft ascended steadily to some 2,000 feet and set its nose eastward for Ireland – nearly 2,000 miles off. And they knew that if they couldn't keep airborne all that way, it was a million to one that they would drown.

Wearing a check cap back to front, Hawker flew straight over the city of St Johns and the Quidi Vidi airfield, where rival airmen were actually preparing to take off on this fantastic race to be the first across the Atlantic and win the *Daily Mail* prize of £10,000.

Hawker signalled 'Farewell' to the crowds gathered below, and then just before leaving the land altogether, he jettisoned his undercarriage, and with it the landing wheels. Many people thought him unwise, but he knew that every extra pound must make the flight more difficult and more dangerous.

The signalman at the marine lookout on Newfoundland reported: 'Atlantic plane flying south-easterly out of sight at 8,000 feet and a speed of 80 mph.'

After only ten minutes in the air, though, Hawker and Grieve lost the strong sunshine as the aeroplane faded into the infamous fog banks off Newfoundland. Hawker decided to try and rise above them, and soon succeeded in raising *Atlantic* well over the swirling sea-mists.

This was better in one way, but worse in another – for the fog blanketed all sight of the sea from them. For over an hour they flew blindly on eastward, though a brief break in the mixture of mist and cloud did give Grieve a chance to take bearings.

For four hours they droned on, then they suddenly ran into clusters of cloudbanks, building up with every eastward mile. Their black outlines looked ominous against the oncoming night.

Four or five hundred miles out, the storm started.

Rain raged against the poor little aeroplane, shaking it from propeller to tail, making every yard more of a struggle both for the men and the machine. The squalls seemed to be coming at them almost horizontally.

The silver wings fluttered, quivered, but remained more or less level. This was the weather they had waited to avoid all those endless days. As night fell they met more great gusts of wind and rain, which roared through the thin struts of the biplane.

It seemed impossible that with everything else coming *down*, the aeroplane could stay *up*.

Grieve tried to take sights with frostbitten hands, as Hawker veered, banked, dived, dodged, to get round those black blots of clouds. Their course became erratic, but they flew on.

Harry Hawker said later:

> 'The trouble did not start until we were five and a half hours out from St Johns.
>
> *'Then the temperature of the water in the radiator began to rise.*
>
> 'It did not mean a great deal at that moment, but we could see that something was the matter with the water circulation.'

They couldn't just stop and cool it off, like a car on a road. They were at 10,000 feet now – two miles over the angry Atlantic. Not many people would be where they were for ten million pounds – let alone ten thousand.

Grieve continued trying to take sights, though they had not seen the water since that moment ten minutes out from St Johns.

Hawker, too, peered outside at the dark facade of sky. He saw cloud peaks towering up to 15,000 feet, making a very bad horizon. The moon had not yet risen, so the whole scene seemed appallingly dark. Yet he said, 'We were very comfortable'!

After the first shock of that rise of the radiator temperature, Hawker glanced down regularly at the dial registering the thermometer level of the cooling system in the engine. At first the water temperature in the radiator read 168 degrees

Fahrenheit. Two or three minutes later it showed 176 degrees Fahrenheit – an eight-degree jump. Things were hotting up.

'Only 36 degrees to boiling point,' he thought aloud.

But luckily there it stayed for the next couple of hours.

Then it started to creep up again.

Hawker wiggled the lever controlling the shutters on the radiator. These were supposed to vary the temperature – but they didn't. The needle showed another two-degree rise.

Hawker had thought it was as much as he could do to fly the plane through the dark drifts of clouds. Now he had the problem of heat to cope with as well. For the first time he felt slightly worried.

If the water reached boiling point and evaporated, their Rolls-Royce engine would overheat and seize up, and they would fall into the hidden storm-ridden waters below.

He started searching for the cause and came to the conclusion that something must have got into the water filter in the feed pipe from the radiator to the water cock and blocked it. Most probably solder or the like shaken loose in the radiator. Then he tried to remedy the fault.

He throttled down the motor, stopped it altogether, and pointed the nose of the *Atlantic* down towards the ocean. He hoped this would give the system a chance to cool, and also clear anything blocking the filter.

The aeroplane drove down through the rain, with the wind twanging the taut wire struts. Hawker screwed up his eyes to try and see something, but couldn't really, so went by his instruments.

Between 6,000 and 7,000 feet he flattened out in a graceful arc and opened the throttle again. He had done the trick. The filter was clear and the thermometer needle flickered back to a lower level.

'Thank God,' he breathed.

After another hour of ploughing through the fleeting films of clouds, they were about 800 miles out from St Johns. The weather was still stormy, with a northerly gale coming up fast.

Then it happened all over again.

The filter choked, the water got hotter; the needle on the dial moved steadily upwards. The moment of truth was approaching, and so was the point of no return.

Hawker did the only thing he could. He dived again to try and clear it, but it was no use this time. And the climb up afterwards heated the water more rapidly still.

The dial now read 200 degrees Fahrenheit: twelve degrees off boiling point.

The aeroplane staggered up though the stormy clouds, the engine getting hotter all the time. The two men were in a tight corner and they knew it. Midnight had come and gone, but for the fliers it was still the middle of a wild night over the Atlantic – in an aircraft that must soon seize up. They were now 900 miles from St Johns, and nearing the point of no return. No men had ever flown here before.

Hawker tried again to clear the filter by diving, but failed. And after the third attempt the water was boiling fiercely.

The dial read over 212 degrees.

If he went on like this, they would waste all their precious water for cooling the engine, so he climbed to 12,000 feet and they decided to stay at that height for the rest of the way.

They hadn't yet seriously doubted the aeroplane's ability to get across. At two and a half miles up, they were above most of the clouds. Moonlight broke the blackness, and they managed to keep a better course. They discovered that the gale had been blowing them southward, miles out of their course.

By throttling down to a slower speed, Hawker got the water off the boil and back to 200 degrees again. Nursing the controls with all the tenderness he could muster, he kept them flying for the rest of that endless night.

Dawn edged alight ahead of them.

Then, twelve hours out from Newfoundland, they came to clouds again, too dense to fly through and too tall to climb over. And anyway, each time Hawker ascended, the water bubbled and boiled.

'Nothing for it,' he shouted. 'Have to go below them.'

But the cloud base seemed to be non-existent. Hawker took

the aeroplane lower and lower. At 6,000 feet the clouds were thicker than ever. Down and down he manoeuvred, praying for a break. At 1,000 feet they suddenly saw the sullen swirling sea. Hawker reacted instantly and opened the throttle.

There was no response.

As the last few hundred feet slipped away, Grieve desperately fiddled with the fuel pump to get the engine going. It looked as if all was lost, when literally yards from the waves, the engine coughed, convulsed, and came back to life.

Up at 1,000 feet again, they were on course as a stormy sunrise loomed red and grey ahead. But even without climbing any higher they could not keep the temperature below boiling. Now they knew the worst – they had passed the point of no return – and they could not hope to reach their goal. It was a grim moment. They could go neither forward nor back.

The water began to boil steadily away, and with it went their hopes. Hawker reckoned he could keep the aeroplane flying for another two or three hours: their expectation of life, for as the Irish coast was still nearly 900 miles off, he knew they could not hope to reach it.

'There's only one chance,' he called to Grieve, 'and that's seeing a ship. We're near the main routes.'

But both of them knew that the chances of spotting a vessel in the vast watery wastes of the Atlantic were extremely small, and to do so before they burned up were virtually nil.

Hawker swung the nose and began to zig-zag across the line of their original route. The minutes, the miles, the fuel and – worst of all – the water, were all being consumed.

Both men craned out of their cramped cockpits, desperately scanning the sea through the scudding clouds. But no ship came in sight. It seemed silly even to hope for one.

This went on for an hour. They had really burned their boats now, and the water went on boiling away. They knew, too, that the chances of setting an aeroplane down on this churning ocean were as slim as sighting a ship.

The *Atlantic* chugged on, back and forth. They had no

radio to call for help. The whole world was waiting to hear news of them, but wouldn't begin worrying till they became overdue in Ireland.

Only these two men in the entire world knew of their plight. All they could do now was to keep their aeroplane airborne for as long as they possibly could. For the throb of the motor was virtually their own heartbeat. When one stopped, the other would soon follow.

It was now two hours since Hawker had decided to adopt their zig-zag course. There was still not a sign of a ship. The temperature needle stayed permanently above boiling point as they flew to and fro.

Hawker still had his hands full controlling the plane, for half a gale was whipping the sea and buffeting the frail fuselage. It was as if they were being softened up for a final knock-out. The rollers rose and fell to the gusts of the gale.

Another half-hour passed, and Hawker wondered how much longer they could last. In his heart he knew it was only a matter of minutes.

Through it all the Rolls-Royce engine purred on, but by now all the water had boiled away. The engine was red-hot and growing hotter.

They were at latitude 50°20′N and longitude 29°30′W. St Johns was now 1,100 miles away, and Ireland 800 miles away.

This was when they should have died – but they didn't.

The miracle happened.

'It's a ship! A ship!' Hawker shouted, pointing down for Grieve to see. They nodded to each other, overcome.

She was quite close on the port bow, and came looming out of the low morning mist.

Hawker swooped down to 400 feet, fired three Very distress signals, and then flew across her until he saw some men on deck. The next problem was *landing* on the sea. At any rate he wouldn't miss his wheels.

He went a couple of miles ahead of the little vessel, veered round, and then came down towards the storm-ruffled surface.

Now Harry Hawker had to call on all his reserves of calmness and skill. He saw the sea spread out beneath him like a rippling landscape. At last came the touchdown. With infinite care he set the machine down on top of the waves – and it actually floated on an even keel.

The next step was for the two of them to take to their little emergency boat. Useful as this was, it would not have supported them very long in the restless rollers of the mid-Atlantic.

The waves were running up to twelve feet and breaking right over them and the aeroplane, which slowly started to settle and sink. Soon only its tail remained above the water.

An hour or so later the ship's boat reached them. The rescuers turned out to be the *Mary*, a small Danish steamer bound for the Scottish coast.

The ironic thing was that this 1,824-ton vessel had no wireless, so the world went on waiting for news of the famous fliers. In fact it was a week after their take-off that the *Mary* passed Lloyd's signal station at Butt of Lewis and broke the breathtaking fact that the aviators were safe, long after they had been given up for lost.

They were both awarded the Air Force Cross, and the *Daily Mail* gave them a consolation prize of £5,000 for their gallant attempt which took them nearly two-thirds across the Atlantic. So the names of Harry Hawker and Commander Mackenzie Grieve went down in aviation annals as the men who were nearly the first to fly the Atlantic.

And the strange postscript: their plane did *not* sink. The US steamer *Lake Charlotteville* spotted it bobbing up and down and salvaged it.

4

COMMANDER READ

across the Atlantic in stages

WHO flew the first heavier than air machine across the Atlantic? Most people would say Alcock and Brown, but they would be wrong. It was Lieutenant-Commander A. C. Read and his crew of the US flying-boat NC-4. Alcock and Brown gained the major honour, of course, by being the first to fly non-stop on a transatlantic trip.

It was the American Navy's flying-boats that really forced Hawker and Grieve into their hurried take-off on 18 May 1919, for no fewer than three US seaplanes had heaved their loads up from Trepassey Bay, Newfoundland, on the afternoon of 16 May in an attempt to fly the North Atlantic to England in stages.

NC-3 and NC-4 took off two minutes apart – at 10.03 and 10.05 Greenwich time – and together left Mistaken Point bound for the Azores on the first leg of their southerly route over the ocean. Ten minutes later they sighted NC-1 several miles behind them and flying higher.

Although their course was in warmer latitudes than Hawker's route, they saw icebergs in the smooth sea below. The whole operation had been planned very thoroughly by the US Navy, which had ships at a whole string of pre-determined points to guide the fliers. The flying-boats had the advantage of being capable of landing safely on the water if the engines failed at any time, provided that the weather was moderately calm.

The seaplanes were each powered by four 400-horsepower Liberty engines. They could fly for twenty hours, so their complete projected journey from Newfoundland to Plymouth would have to be carried out in three stages: Trepassey to the

Azores, 1,381 miles; Azores to Lisbon, 1,094 miles; and Lisbon to Plymouth, 895 miles. The total distance therefore amounted to 3,370 miles. NC-4 was the newest of the flying-boats, with a take-off weight of 22,000 pounds.

They flew at about 800 feet. NC-4 drew ahead but when it reached the first US destroyer it circled around for NC-3 to catch up. Then for the next hour they flew on together, until the running lights of NC-3 gradually grew too dim to be distinguished and NC-4 lost sight of it altogether. From then on NC-4 proceeded as if alone, with the engines turning well, and both oil and water temperatures correct.

That night their only light at first was the stars showing in a dark sky, but then at midnight the May moon began to rise, somehow comforting the crew of NC-4. The air became bumpy, so they climbed to 1,800 feet, but it was no smoother there.

They sighted each destroyer in turn in the positions appointed for them. Star shells, seen from up to forty miles away, first signalled the ship's position, then searchlights and ship's lights guided NC-4 on its well-marked route. All the vessels were brilliantly illuminated; some ships were precisely in position, others less exactly.

So far their average speed had been 104 miles an hour, indicating a 14 mile-an-hour favourable wind. At 5.45 am they saw the first smudge of dawn. As it spread into daylight, most of their worries seemed to be past, at any rate for the first leg of the flight: they were half way; the power plant and everything else was functioning perfectly; the radio worked wonderfully. The radio officer had even sent a message to his mother in the United States via Cape Race, then 730 miles off. Messages came in from as far as 1,300 miles.

Cape Race reported: 'NC-3's radio working poorly.'

By intercepting signals they learned that NC-3 flew ahead of NC-1 and both were astern of NC-4.

The crew of NC-4 had a comparatively comfortable time, eating sandwiches and chocolate candy and drinking coffee from the thermos bottles. Lieutenant Commander Read made several inspection trips aft and asked the radio operator and the engineer how things were going.

At 6.55 am they passed over a merchant ship and at 8 am the flying-boat coasted through light patches of fog. This soon passed but at 9.27 am they met more fog – dense this time.

The sun faded from an orange outline to nothing at all and they rapidly lost all sense of direction. The spinning compass suggested a steep bank and Read had visions of a possible nose dive. Even the best precautions could not have avoided this. But fortunately it did not last long and the sun suddenly pierced through again and the sky became blue instead of foggy-grey. Read put NC-4 on an even keel and climbed above the fog and upper limit of the layer of clouds. Their altitude now reached 3,200 feet, still not high though substantial for the weight being borne. The fog remained below them.

At 10.38 am and 10.55 am they sent out radio messages to the nearest destroyer, asking if the fog had lifted near the ship. But the replies were not encouraging, reporting thick fog near the water. Light rain brushed the flying-boat but passed like a spring shower. Then at 11.13 am they radioed the destroyer *Corvo* asking about conditions further on and heard the reply: 'Visibility ten miles.'

Cheered up by this promise of better, brighter conditions ahead, NC-4 thundered on for the Azores.

Suddenly at 11.27 am a hole in the clouds and fog gave them a glimpse of the sea – and what looked like a tide rift on the water. Two minutes later they saw the ragged, irregular outline of rocks. The tide rift was in fact a line of surf along the southern end of Flores Island, the most westerly of the Azores.

'It was the most welcome sight we had ever seen,' said Read afterwards.

This told them, too, that they were forty-five miles off their calculated position, indicating that their speed from the last destroyer sighted had been about 97 miles an hour. The wind was blowing them east and south. They took their bearings from the island and shaped their course for the next destroyer, flying low with a strong following wind. The fog stopped 200 feet above the water.

At noon they looked down on destroyer No. 22 right in her proper place, the first in the chain of destroyers they had seen since No. 16. As visibility had then extended to twelve miles, and they had plenty of petrol and oil, Read decided to keep on for Ponta Delgada, their real goal in the Azores.

But then the fog swirled round them once more, and they completely missed the next destroyer, No. 23. Then the fog really closed down, but they decided to keep to their course until 1.18 pm and then make a ninety-degree turn to the right to pick up Fayal or Pico island. However, at 1.04 pm they made out the northern tip of Fayal and felt safe again.

They headed for the shore, the air clearing quickly as they got nearer to the beach. NC-4 rounded the island and landed in a bight they mistook for Horta. The error did not matter much and they simply took off again, leaving a trail of foam, and rounded the next point. Through the fog the dark blur of a vessel clarified into the USS *Columbia* and NC-4 landed near her at 1.23 pm.

The total time since take-off had been 15 hours 18 minutes, at an average speed of 94 miles an hour.

NC-4 left Horta for Ponta Delgada to take in fuel, and they hoped to make only a brief stay at the island of San Miguel before aiming due eastward for the next destination – Lisbon.

While NC-4 waited to continue its journey, attention naturally switched to the other two flying-boats behind it – NC-3 and NC-1.

The crew of NC-3 under Commander Towers had a terrifying experience. The failure of the lights on the pilot's instrument panel forced him to fly above the clouds to see the stars for guidance. The last destroyer he sighted was No. 12. He dropped below the heavy clouds at daybreak but missed destroyer No. 14. He assumed that the high velocity of the upper winds had hurled him off course. He laid a parallel course, but at 7.45 am GMT on Saturday, 17 May, he ran into heavy rain squalls. These went on for nearly six hours, when the weather cleared.

Commander Towers said afterwards: 'We then decided to land to make observations, as we had only two hours' fuel left. We discovered a heavy sea running, too late to remain in the air.'

When NC-3 alighted on the water, the impact slightly damaged the hull and seriously damaged the forward centre engine struts. They could not take off again – and they were still just over 200 miles from Ponta Delgada.

A gale sprang up that evening, but the flying-boat rode out the stormy night. The seaplane suffered severely but the crew succeeded in keeping it afloat all through the gale. At 9 am the next morning they lost the port pontoon.

They had to try for Ponta Delgada, but they could not fly, so the only way to get there was on the surface of the sea. Slowly, bumpily, NC-3 taxied the 205 miles from the point where it came down right to Ponta Delgada. It arrived there at 5.50 pm on 19 May, two days later, under its own power. They lost the starboard pontoon just outside the harbour, but they were there – safe.

The fate of the NC-1 and its crew was even more remarkable.

Like the NC-4 and NC-3, NC-1 had left Newfoundland just after 10 am GMT on 17 May. Its flight was more or less similar to that of the other two for the rest of that day and night.

With the glow of dawn, everyone in the crew felt confident that they could make Ponta Delgada easily. Then they began to run into thick overcast patches and visibility deteriorated quite dramatically. As they went through one patch, Station Ship No. 16 loomed dead ahead of them. Some of the station ships radioed weather reports to them. They passed No. 17 on the port hand at a range of twelve miles.

They flew on at 600 feet, and then ran into really thick fog. It dimmed the goggles of the crew and misted the glass over their instruments, so that they could scarcely read the dials.

The pilots, Barin and Mitscher, brought the seaplane up to 3,000 feet, above the fog, but at this height they could not see the water and so did not know how far they were drifting off course.

They dodged one patch of fog but kept running into more. They side-slipped and turned in an effort to keep on course until at 12.50 pm they decided to come down, get their bearings and fly underneath the ceiling.

At seventy-five feet over the water, visibility extended to half a mile, the air was bumpy, and the wind shifted from 350 to 290 magnetic. They changed course to conform to the fresh conditions and sent out radio signals requesting compass bearings.

They decided to land if the fog thickened. It did.

A few minutes afterwards, they ran into a low, frightening fog-belt. Lieutenant-Commander P. N. L. Bellinger turned the flying-boat about and headed into the wind, landing at 1.10 pm after a flying time of fifteen hours.

The water was much too rough for them to try and take off again. The outlook seemed gloomy, even grim. The wind and water both prevented the plane from taxiing over the surface to windward, as NC-3 had done, and they soon found that radio contact between the aeroplane and the ships was erratic and unsatisfactory. They realized they had to wait where they were and hope to be found.

Shortly after they touched down they put over the sea anchor, but the strong swell carried it away almost immediately. Then they rigged up a metal bucket as a sea anchor, which survived and did a lot of good.

The wings and tail, however, went on receiving pounding punishment from the rough sea, so they slit the fabric on the outer and lower wings to help preserve the structure. In a further effort to reduce the damage Bellinger kept one of the centre motors running.

Despite all their efforts, the weather badly buffeted and damaged both wings and the tail. For some time it seemed as if NC-1 would actually capsize, although it was a seaplane and so designed to float. All the men realized the danger, but showed no fear. They waited for hour after hour.

Finally at 5.40 pm someone said: 'Look! There's a steamer!'

The seaplane taxied across towards her. The vessel was the *Ionia,* carrying no wireless. She sighted the Americans but before she could get over to them, the fog clamped a curtain of grey between them, and the seaplane vanished. Later the weather lifted and the aircrew saw the ship steaming towards them. The *Ionia* took them in tow. A destroyer came alongside during the night and took charge of the battered NC-1, while the crew were later landed safely at Horta in the Azores from the *Ionia.*

Three seaplanes had set out – but now there was only one.

NC-4 took off from Ponta Delgada at 10.18 a.m. on 27 May on the second stage of its flight, from the Azores to Lisbon. Commanding it was the same small, lean, wiry thirty-two-year-old man, Lieutenant-Commander Arthur Read. They had 1,094 miles to fly before they could claim to have crossed the Atlantic.

The sea was speckled with waves, and the flying-boat had a good following wind. Thick clouds draped the hills of San Miguel as NC-4 gained height to 800 feet.

The Americans had fourteen marker boats out, and the naval seaplane saw them all but one. The whole way to Lisbon was without adverse incident. When they could not spot a ship, they could contact her by radio, so communications remained complete despite the mist which still seemed to be dogging them.

The grey body and yellow wings of NC-4 thrust ever eastward, its high-pitched tail planes detracting from any appearance of grace. No one could really call it beautiful, though that is what it seemed to Read.

The navigator plotted the course, seated in a cockpit forward in front of the four engines, while the two pilots sat side by side just behind the engines. The engineers were aft with the radio operator.

So the flight went on and they had little to report till they reached Lisbon in 9 hours 25 minutes, to a great welcome. They had flown the Atlantic.

The speed of NC-4 on the second leg varied from 69 to 104 miles an hour depending on the winds, but not once did it depart far from its direct course to the Portuguese capital. The average speed for the thousand-odd mile hop was 93.7 miles an hour.

So to the final flight: a run of 895 miles north-north-east from Lisbon to Plymouth. NC-4 rose from Lisbon harbour at 5.29 am GMT on 30 May with a favourable wind behind it but amid squalls. At 7.05 am they discovered a leak in the port engine, so descended for repairs at the mouth of the Mondego river.

From there NC-4 left at 1.38 pm for Ferrol harbour in Spain, which it reached at 4.47 pm. After an overnight halt to give the crew a badly needed rest, they left Ferrol at 6.27 am the next day, 31 May, sighting only two destroyers en route to England, due to the persistent squalls. They reached Brest about 11 am in showery squalls, typical summer weather for Europe. A headwind did not deter the Americans from the last lap of all – across the English Channel.

Plymouth waited while those two final hours elapsed, and at 1.12 pm the crowds spotted the seaplane east of the Sound.

Very lights, sirens, and all the other paraphernalia of welcome made it clear that Britain regarded this as a historic moment in the annals of flying, which it was. The next one came only a fortnight later.

5

ALCOCK AND BROWN

the Atlantic - non-stop

JUST under four weeks after Harry Hawker and Mackenzie Grieve had taken off from Newfoundland, Alcock and Brown started on their attempt at the first Atlantic flight. Like Hawker, both Captain John Alcock and Lieutenant Arthur Whitten Brown had gained their experience in the Great War, and in fact they chose a modified bomber for their flight.

This was the Vickers Vimy, a large biplane bomber carrying 865 gallons of petrol, enough to supply the two 375-horsepower Rolls-Royce engines for a journey of 2,440 miles. The actual point-to-point distance between Newfoundland and their Irish destination amounted to about 1,890 miles, so they had fuel for 500 extra miles in reserve. The machine measured forty-three feet and had a wing span of some sixty-nine feet.

Profiting from Hawker's experience, they made certain that the Vimy's water system was working perfectly, and all the water to be used was first filtered, boiled and then strained before entering the tanks. If either of the two engines were to fail after the halfway point, they could carry on to Ireland on the remaining one. If this happened before they had consumed half the fuel, however, the other engine could not support such a load and they would be forced down into the sea.

Alcock and Brown carried Very lights to help meet any emergency like this, and they also had the great advantage of wireless communication. Altogether they seemed to be better prepared than Hawker and Grieve, wearing electrically heated clothing and taking plenty of nourishment, including special sandwiches, chocolate and hot drinks.

The two men sat side by side, but Alcock was to be at the wheel all the way. The Vimy did actually have a driving 'wheel'. Brown said before the flight: 'I suppose I shall be butler and everything, for Alcock will be too busy flying the machine to get the food out.'

It was lucky that the pair were not superstitious, for the date when they finally felt ready was Friday, 13 June 1919!

Alcock's words before the take-off the following day were: 'It's a long flight, but it doesn't worry me any more than the night bombing raids my squadron used to carry out.'

Over their last hurried meal before leaving, Brown added:

> 'With this wind we shall be in Ireland in 12 hours. The air speed indicator shows a ground speed of 42 miles an hour. We can't hope for that all the way, but the winds are favourable, and we might be in sight of land in anything between 19 and 20 hours. I am steering a straight line for Galway Bay.'

Brown intended to use the ordinary Mercator chart of the North Atlantic and reckoned to find the position of the aeroplane at any time by observing the height of the sun, or a suitable star, through a sextant, noting the Greenwich time and plotting the result by a special chart showing the curves of equal altitude. He would also take account of the effect of the wind on them by a drift-bearing plate.

The fuel weighed 3½ tons, and the whole aircraft over 6 tons. Despite the 40-mile-an-hour gale gusting into the airfield, Alcock decided to go ahead. It was 4.13 pm GMT on that Saturday afternoon when they started up the two engines. After Alcock had listened to them for a few minutes, he felt satisfied with their behaviour and gave the signal to clear the chocks from the wheels.

The long tapering body of the Vimy vibrated to the gale and the pulsing purr of the engines, like an animal eager to leap. Alcock took the aircraft forward uphill – the field had a fairly steep gradient.

It took nearly a quarter of a mile before the plane eventually left the ground. Nearly all the population of St Johns had climbed to Lester's Field to see the start. Rising steadily the Vimy passed over the town and across the White Hills, at the point where Hawker had set out nearly a month earlier.

Then the crowd had a shock as it appeared that the plane had dipped and disappeared, while it actually flew up a valley. They had to open the engines full out to get up that valley, thrusting the whole power of the Vimy against the force of the funnelled wind.

They crossed Signal Hill, Newfoundland, at precisely 4.28 pm GMT, the official time of the start of the flight, and almost from that moment on things began to go badly. It looked like being Hawker and Grieve's experience all over again. The weather was certainly similar.

They put themselves at the mercy, or lack of it, of the elements – fog, ice and wind. In fact, despite all the planning, Brown was able to take only four readings of their position: one from the sun, one from the moon, one from the Pole Star, and one from the star Vega.

They quickly climbed to the agreed initial height of 1,000 feet and at Signal Hill set out a course for the ocean on 124 degrees compass point. The elements then almost took over completely: the strong south-west wind added to their speed, but the fog and clouds closed in all around, blinding them.

Early on Alcock throttled down nicely and let the plane gradually climb. At dusk they had got up to 4,000 feet, and found themselves sandwiched between layers of clouds, and unable to see either sea or sky. The lower level of fog and clouds drifted at about 2,000 feet and the upper one at 6,000 feet. Before dark the clouds had hidden the sun altogether. They could not see the sea, and when night came the clouds concealed the moon and the stars. It was all pretty comfort-less.

Those first six or seven hours were really bad. The fog

hung low over the water, the clouds moved north-east before the wind in the same general direction as the aeroplane. Alcock could only go on trying to fly blindly between the main layers above and below.

Alcock divided his time between straining to see ahead and watching the air-speed indicator and compass. In this way he hoped to keep the Vimy roughly the right way up.

Still struggling forward on their original course, they came to a momentary magic break in the conjunction of cloud and endless, enveloping, Atlantic fog all about them.

Through the crack in the clouds, they saw stars sparkling and twinkling. Alcock took the bomber up towards the gap and called to Brown: 'Now's your chance. Can you get a bearing?'

He could and did, getting a 'cut' on Polaris and Vega as Alcock managed to keep the aeroplane on a level keel with the unaccustomed luxury of an actual horizon ahead of him. Rapidly Brown did his calculations and gauged that they were about two degrees south of the course they had set.

'New course 110 degrees compass point,' Brown told Alcock, who adjusted the plane accordingly and flew on.

The clearance of clouds turned out to be brief, so it was lucky they got their bearings when they did. Soon the weather started to thicken again. They were at those darkest hours before dawn, when flying in a clear sky would have been hard enough. But to control a big bomber as it throbbed through blanketing banks of fog was almost too much for Alcock.

'We began to have a very rough time,' admitted Alcock later.

He could not see the sea, the sky, or where they met. The only thing he could rely on was his air-speed indicator. Then it jammed.

Sleet had been falling for the last hour or so and in icy conditions nearly a mile up in the Atlantic night, the sleet had frozen on the indicator. They could also smell smoke.

So whether Alcock opened the throttle or closed it, the air-speed indicator still showed the same reading: 90 miles an hour. Only this device could have told them that they were flying roughly level, for if they rose or fell in altitude, their speed would change provided the throttle remained the same.

Its failure would not have mattered so much in daylight or on a clear night. But to be fogbound, cloudbound, sleet-bound, and not know whether they were flying level or not – that could quickly be fatal. Alcock knew it. He concentrated all his supreme skill on keeping the aircraft as steady as he could, and for a time the engines responded with a normal note, but not for very long. It was surprising that he had managed it at all.

It was a terrible sensation, that utter lack of a sense of horizon or gravity, like floating around in a fourth dimension. But they were still very much in a world of three dimensions and if they were not careful they would be in the ocean, drowning.

Now the engine started to shift from the Rolls-Royce whirr to a protesting whine and roar.

'We did some comic stunts,' said Alcock afterwards. 'It was very alarming.'

Through the thick clouds they actually looped the loop in a frantic effort to find some sort of stability. But it was no use. From the loop and ensuing slight climb, the Vimy went into a ponderous but quite steep spiral. Alcock couldn't correct it. He hardly knew what to do. He waited for a few seconds, nerves strung as taut as the wires between the struts.

From 4,000 feet the aeroplane rolled round and down, gathering speed as it fell further through the spiral. Alcock was literally helpless now. It all happened so quickly. He couldn't fight gravity when he still scarcely knew which way it was pulling.

All this took only seconds. The Vimy roared round and round. Their altitude fell to 1,000 feet, with visibility still virtually nil. All Alcock could hope for was to hang on to the wheel and wait.

It seemed impossible that Alcock could ever regain control. Suddenly they came out of the clouds and Alcock spotted the dim suggestion of the sea. It was enough. A faint blur, but a horizon of sorts. He had to act instantly, for they were down to 200 feet already, and still spinning.

Using all his wits and strength, he began to pull the aeroplane out of that alarming angle, but it was already almost on its back.

At a hundred feet it was under control again, and at fifty feet it had levelled off.

So they were saved, less than a wing span above the water. It was the closest call Alcock had ever experienced. Only by his instinctive reaction did he manage to right the fall of the heavy aeroplane.

That fall could easily have been their last. It did have one beneficial effect, though, for the air-speed indicator began to work properly again after the shock treatment of that swift dive.

Gradually Alcock and Brown got over the fright of those few seconds. The wind remained more or less behind them. They climbed to 6,000 feet, then they ran into the fog again, even at that altitude. To look into it for long hypnotized the mind and made the eyes watery.

Alcock climbed above the fog twice, only to run into congestions of clouds each time. And in one area one bank of fog lay actually on top of another, lower one. The aeroplane emerged from the gloom once, and the pilot snatched a peep at the moon and surrounding stars. But before Brown could do much about fixing their position, the aeroplane ran straight into a vicious combination of hail, snow and sleet at 11,000 feet.

As far as navigation went, they were trusting to the compass. In any case they had far worse things to worry about as the night wore on than flying off course.

Ice threatened to envelop the whole aircraft. The radiator shutter and water temperature indicator became covered in it, and it put the air-speed indicator out of service once more. They were still amid the fog and cloud. Alcock said suddenly, 'It's jammed the ailerons.'

Alcock knew that the presence of ice on the radiator shutters would sooner or later react on the engine. It seemed as if something must give out before they reached Ireland. However, the most pressing problem was the air-speed indicator, so Brown forced himself up from the cockpit to a position near the centre-section strut. Exposed in this way, the navigator risked falling out of the machine altogether, but he somehow struggled on, to chip off the ice with a knife. This improved the air-speed indicator, but he had to repeat the operation regularly. It gave trouble again later, when it became full of frozen particles.

They never saw the sun rise, but finally after some thirteen hours, the clouds did scatter slightly to let in dim daylight. The wild weather remained with them and so did the ice.

Two or three hours from the Irish coast, Brown took some rough readings, but could not tell where they were with any certainty, so he motioned for Alcock to go down lower, below the clouds. They had communicated by shouts or gestures most of the way, for their intercom telephones had broken down a quarter of the way across. And as for wireless messages, they never heard so much as a single one!

The pilot tilted the nose of the Vimy and there was almost a repetition of the near-disaster back in mid-ocean. Down and down they dived, till they shot out of the low clouds. Alcock was able to straighten up with comparative ease and he kept the Vimy throbbing and thrumming eastward as Brown tried to take any checks he could.

They flew over the sea at a mere 300 feet, as the sun made a fitful, feeble attempt to appear, but was defeated by the miles of clustering clouds.

'It was a terrible trip,' Alcock said afterwards.

Only after another hour at this low level did the frozen particles start to fall off the air-speed indicator, but they did not really need this much now. They followed the horizon at a height of a mere hundred yards. They were not sure where they were, but by time and indicated speed, it seemed that they must soon see land. Any part of the cliff-lined coastline of Ireland would look like paradise itself, but still the water spread out ahead, a mixture of cloud-grey and sea-green.

Those last minutes seemed longer than the rest of the flight put together. Tiredness crept over them as they looked alternately at the instrument panel and the horizon.

'Should be roughly on course,' Brown called above the wind, after another careful check.

They had flown and fought the elements for nearly 1,900 miles. Where was the land?

Then they saw two little islands, Eashal and Turbot. The two men spotted them simultaneously, and a few miles further on they saw the Irish coast itself, at about 8.15 GMT on Sunday, 15 June.

In no time they were over land and looking down on Ardbear Bay, an inlet of Clifden Bay. When they saw the slender mast of the Clifden wireless station, they knew exactly where they were. They had navigated to the precise point they had aimed at – and there it rose below them in the grey morning. Alcock and Brown exchanged quick glances. They had no need for words.

Alcock circled over the village of Clifden, craning out of the cockpit to spot the likeliest landing place. He saw what looked like a lovely green meadow, and banked to bring the Vimy round for the historic landing. Throttling down, he idled the great Rolls-Royce engines, and positioned for a perfect run-in.

But the meadow was a bog! The moment the four wheels touched down, they started to sink axle deep into the soft mud. The inevitable occurred. The Vimy toppled on to her nose; the tail tipped up at forty-five degrees; the lower wing crunched and crumpled; the propellers dug into the bog; and Alcock and Brown were pitched forwards.

But they weren't hurt and hoisted themselves out to realize that they had become the first men to fly the Atlantic non-stop. They had done it in 16 hours 12 minutes.

A big broad smile covered John Alcock's ruddy face.

Arthur Brown smiled, too, looking slightly more tired. Both men were knighted and received the £10,000 prize from the *Daily Mail.*

Tragically Alcock was killed just six months later on a routine flight from England to France.

6

ALAN COBHAM

an epic flight to Capetown

ONE of the greatest 'first flights' of all was Alan Cobham's 17,000-mile epic from London to Capetown and back: only surpassed by his amazing Australia-and-back venture immediately afterwards.

Alan Cobham took with him on the Capetown trip two other men, A. B. Elliott as engineer, and B. W. G. Emmott, a Gaumont-British photographer who was to make a film of the whole flight.

Their plane was a De Havilland Type 50 biplane. This had a passenger cabin to the rear of the engine, with the pilot's own cockpit situated separately, though Cobham was able to talk to the others. The plane was powered by a 385-horsepower Siddeley-Jaguar engine, strong enough to convey the mass of gear for such a stupendous undertaking.

The three men took off from Stag Lane aerodrome, near Hendon, on 16 November 1925, but only got as far as Paris that day. Next day they flew on to Marseilles, making another overnight stop.

The weather was bad by now, and a surging north-easterly gale swept up and over the southerly Alps. In an attempt to escape the violent air turbulence, Cobham took it up to 6,000 feet, but the turbulence got even worse. He appreciated that he dare not risk these conditions for long, or they would fail before getting further than Europe. So he swung the aeroplane out to sea, and skimmed over the choppy Mediterranean towards Pisa and the leaning tower that gave them their bearings.

The next stop was Taranto, at the heel of Italy, and from there they hopped across a stormy sea to Greece, again finding themselves buffeted by wild winds gusting over mountains. Cobham came down low over the water once more, crossing the Gulf of Corinth and making Athens by 20 November. The 480-mile flight south over the Mediterranean to Africa was uneventful and they reached Cairo safely and settled in there for a week or so, making several flights over the Pyramids for Emmott to take pictures – quite a novelty at the time.

They resumed their journey and flew on down the course of the Nile over Luxor – more opportunities for filming, this time the Aswan Dam. They were aware that they were witnessing what no one had seen from the air before, the ever-shifting ageless panorama of Africa. The flight became a list of exotic place-names like Wadi Halfa, Atbara, Khartoum. Before getting to this outpost of Kitchener's days, they saw beneath them the junction of the Blue and White Niles.

They spent Christmas at Khartoum, having flown all the way from London. In 1925 that still seemed fantastic, like a Wellsian dream. But to Cobham, it conveyed a glimpse of a realizable future for international airlines.

At Malakal they watched a war dance by the local Shulluk warriors, who even posed before the aeroplane. The contrast between these primitive people, who had never seen an aeroplane, and the machine itself was quite dramatic. They stood with simple dignity, even indifference, before the DH-50 with its markings 'Imperial Airways Air Route Survey' painted beside the enclosed cabin. Then they moved to the beat of tom-toms, brandishing spears and disturbing the dust. It was all extremely strange.

So far the plane had functioned perfectly and Cobham foresaw no problems. The next stage in this pioneering flight was to Mongalla. The direct line lay over the swampy Sud region, which would prove dangerous and even fatal if they were forced to land. Cobham kept to the east of the Nile, which slushed a course through the Sud, and in this way managed to skirt the worst areas of this unknown swampland.

The next point was Jinja, almost on the equator. This place, beside the Victoria Nyanza lake, was several thousand feet higher than the aerodromes Cobham had negotiated so far in Africa – and he nearly had an accident because of this change.

The colourfully clothed natives of Uganda practically filled the whole runway at Jinja as Cobham brought the biplane in to land among them. He was piloting the plane just above a banana plantation, on the edge of the aerodrome, when some of the locals got so excited they ran right across the runway in the path of the machine.

Instinctively, Alan Cobham reacted as he would have done in such an emergency in England and started to land slightly short of them, hoping in this way to avoid any chance of hitting them.

Coming down shorter meant slower, too, but in the rarer air of Uganda, 4,000 feet above sea level, he needed a higher landing speed than he possessed to keep the plane airborne until the actual touch-down.

The De Havilland plummeted no less than ten feet on to the baked Uganda ground and the three men received a sharp jolt. Luckily the undercarriage was equal to the force of impact and did not show signs of damage. Cobham had learned his lesson, however, and did not forget it for the rest of the flight.

At Kisumu, an elderly lady asked them, 'How do you manage to sleep at night?' thinking that they spent several days at a time in the air. She seemed surprised when they told her they landed.

After Tabora in Tanganyika they ran into the rains with a vengeance. The date by then was 18 January 1926.

Cobham found this sector, from the southerly tip of Lake Victoria Nyanza to Palapye Road in Bechuanaland, about the worst of the entire trip. They had to travel over total jungle all the way, with no chance of setting an aeroplane down in the event of engine failure. Cobham knew nothing about what sort of weather to expect, apart from general indications.

They flew on to Abercorn, Northern Rhodesia, and N'Dola, above the vast Lake Bangweolo. The railway line guided them at this stage. The landing grounds here, as at other similar spots, were two runways in the shape of a cross, so that Cobham could come in from any one of four directions, depending on the wind. Knowledge of such things as prevailing winds was sketchy at that time, to say the least.

The next stage of their journey was from N'Dola to Broken Hill, over the railway, and on towards Livingstone, named of course after the explorer. The little wings cast their still smaller shadow over the trees and scrub and railway.

They left Broken Hill soon after breakfast, bound on the 290-mile hop to Livingstone, still following the rails. After a few hours' flying, and with virtually all of that stage behind them, they saw smoke, or what they took to be smoke, on the horizon. Suddenly Cobham realized it was not smoke but *spray*. They were flying towards the legendary Victoria Falls and nearly never flew away from them.

Livingstone spread away beneath them, but they were attracted to the Falls. Cobham flew on the few miles to the scene. They gasped as they gaped down from the cabin at the mile-wide Zambezi. Cobham kept to his course and was soon over the top of the waterfall, that mile and a quarter width of water eternally rolling, roaring, falling beyond the brink and converging on the single sheer chasm.

The Falls suck masses of warm air down the gorge with the water, and this rises later, sending up the spray in liquid mist to 1,000 feet higher than the lip of the falls.

After the two others had taken moving and still photographs, Cobham flew them in close along the line of the brink, for Emmott to film the whole vast expanse of water as it hurled itself over the top. Cobham decided to fly as slowly as he could, without undue risk of stalling.

After the approach at the westward end, he started along the actual brink, fifty feet above it and fifty yards away from it. He hoped to make his pass last a couple of minutes or so.

Suddenly the spray surged briefly up and around them. Globules of water dripped off the wings.

It happened again, and they were lost for a few moments in a fog of fine spray. The plane bumped about. They came out of it all right and headed more steadily for the far side of the Falls, with less than half a mile to go.

Nearly at the end of that great gaping brink, the De Havilland vanished into a thick vaporous spray, just as they were above the chasm.

Cobham was worried. He could not see a thing, but he knew that only feet away on their right rose the gigantic railway suspension bridge spanning the gorge. On the port side, the millions of gallons of the Zambezi went on thundering down. And ahead were rocks and jungle.

Then the engine croaked.

They had flown all the way from London to the Victoria Falls and it had not happened before. It stopped, started, stopped again.

Cobham turned to look at the other two through the dividing window. The trouble was that the air-intake pipes to the carburettor were sucking in spray. Cobham had to try to keep the engine going and the propeller revolving through the fine fog.

The engine went on faltering.

Luckily the engine just managed to fight the spray in the pipes, but Cobham still had to get clear of the Falls. Another serious engine cut-out and they could still speed down into the churning torrent below them.

He hauled back the control lever, and their reserve of power enabled the aeroplane to climb. Continuing to keep the engine full out, Cobham climbed as fast and as far as possible.

The aeroplane responded and rose several hundred feet over the Falls before the pilot started to veer off towards the airfield. They were not out of danger yet, he knew, for the engine was still spluttering. Cobham had to try and get completely clear of that area. It was no use being caught with an engine failure over the forest; that would be as bad as the Falls.

He willed the aeroplane away from that petrifying spray and the great jaws of the gorge. The engine reacted marvellously and bore them back towards the landing ground at Livingstone. Only when Cobham was sure that he could glide down did he feel safe. It had been touch and go. They realized this more fully that afternoon when they visited the foot of the Falls at ground level and were suddenly soaked by the fierce spray storm.

The rest of the route was quite tame after that: Bulawayo, Pretoria, Johannesburg, Kimberley, Bloemfontein, Beaufort West, and finally Capetown on 17 February, three months, and 8,500 miles, since that foggy day at Stag Lane.

Alan Cobham had made air history in those months but now he wanted to improve on the time taken – to bring Britain and South Africa closer together. He had another incentive to get home quickly, too, for on 26 February, the very day they took off again, the steamship *Windsor Castle* left for Southampton, and a unique race was launched between the liner, which would sail 5,300 miles, and the aircraft, due to cover 8,500 miles.

At the Livingstone–Broken Hill stage they had trouble, but not so serious this time. Cobham had to fly through hours of misty rain over hundreds of miles of utterly uncharted interior. If they came down anywhere there no one would know.

At N'Dola five inches of rain in four hours turned the landing ground into a morass of mud. That put them back a day.

While they were landing at Tabora, one wheel of the plane sank up to its axle into the ground, and it took quite a force of helpers to pull them out. The take-off was worse. As Cobham taxied towards the take-off point, one wheel or the other sank into the soft earth every yard or two. At last they did decide to go ahead and try to get up, but the wheels clogged and clamped them down. All the while it seemed as if the De Havilland must tilt forward, but at the fringe of the field it managed to heave itself a few feet off the ground at a speed scarcely faster than a car. It was one of the nasty moments of the return route.

The next headache proved to be the torrid tropical heat. It was tough on the men but worse for the machine and gear. The lubricating oil started to warm up alarmingly, but they managed to maintain oil pressure. Cobham found that at times he could not look out over the fuselage, as the draught scorched past like an actual flame. Landing was like entering an oven, so airless did the atmosphere seem at ground level.

Cobham and Emmott were doing rather more than Elliott, who had suffered an attack of malaria on the outward stop at Johannesburg and was still convalescing.

In some respects, the return journey was worse than the outward, in spite of the fact that it was so much shorter. They took off from Khartoum amid the after-effects of a sandstorm, and even at heights of 10,000 feet and beyond they still had to fly through clusters of the swirled sand. Visibility was just a word.

Coming down carefully through the sickly-yellow, sand-laden air, they reached Atbara, to be told that the dust was thicker still ahead. But they risked it. Climbing once more to nearly 12,000 feet, Cobham began to find it more and more difficult to follow the course of the Nile, his only reliable guide. The thought of a forced landing in the desolation of the desert haunted him.

Then he suddenly said, 'It's *not* the Nile!'

He realized they were lost in the banks of dust between aeroplane and ground. What he thought had been a river was not. Cobham came down through the two miles and more of dim dusty air, the propeller threshing through the thousands of minute particles. Nothing on the ground gave him a clue as to their position.

Then he saw an old dried-up waterway, followed it, and found the Nile. That extinct river bed may well have saved their lives. He then flew at a height of twenty feet to Wadi Halfa.

After leaving the Nile, Cobham followed the telegraph poles projecting above the straight railway line, keeping as close to it as he could. Elliott was busy checking the engine records. They got in to Wadi Halfa, filled up their tanks, and nosed on for Aswan, which they reached at sundown – glad to be back to comparative civilization.

Reaching Egypt on 7 March they set up the record of the first Capetown–Cairo flight.

The tail skid broke while they tried to take off from Sollum for Athens, which delayed them for a day. With an extra-heavy fuel load, they struck north over the Mediterranean for Greece. Cobham reckoned to pass over Crete, but even if he missed that island, he estimated it to be only a two-hour hop to the Greek coast.

But visibility shut down sharply to a mile. After three hours he still saw no land. Three and a half hours passed and still nothing. Rain streamed off the windscreen, reducing the outlook to a grey blur.

Cobham began to wonder if he had missed Crete and Greece, and was heading for the open Ionian Sea. Their fuel supply drained away with every mile they battled through the murk. Half a dozen times they thought they saw land; then at last they really did. It was not Crete, but the isle of Kythera. Soon the mainland loomed on a dim horizon.

Cobham struck down-draughts again off the mountains. The powerful currents bumped and bounced them all over the sky. Then one tremendous thrust forced the three men and all the gear right up against the roof of the cabin. The petrol in the top tank was poured up against the roof. The fuel was gravity fed, but they were falling faster than gravity could operate, so no fuel got through to the engine and it stopped.

As the aeroplane was pushed towards the sea, it met a sharp up-thrust, the engine began to get fuel, and it re-started. They reached London a couple of days later on 13 March 1926, after a fifteen-day flight from Capetown. They had beaten the *Windsor Castle.*

7

ALAN COBHAM

to Australia by seaplane

FROM England to Australia and back in a seaplane. This was what Alan Cobham undertook only three and a half months after the triumphant trip to Capetown.

With his mechanic, A. B. Elliott, he took off from the slipway of Short Bros works at Rochester in Kent, before 5 am on 30 June 1926. The De Havilland Type 50 biplane had been fitted with seaplane floats instead of the normal undercarriage, and ahead of it lay a two-way trip totalling 28,000 miles. Apart from four test flights the week before, Cobham had never flown a seaplane at all! He did know a lot of the route, however, as he had already made a journey to Rangoon and back.

The maximum load for the plane was 4,200 pounds, but they were actually bearing about 5,000 pounds. Despite this extra loading, the Siddeley-Jaguar engine seemed able to cope quite well, and before long they were cruising across France at 100 miles an hour. Cobham had only rather a crude means of working out if he had enough petrol to take them to any particular destination. It went like this: 'If I can do 220 miles in so many minutes, how long will it take to do 670 miles, etc.'

His arithmetic proved accurate and they came down on the water by Marseilles after a non-stop hop of 6 hours and 40 minutes. Cobham liked the freedom offered by a seaplane: wherever water existed, he could land. That was the theory.

During the next jump, across to Naples, he made rough measurements of the distance flown by using his fingers on the map! Gliding gracefully into Naples Bay, he had a nasty jolt. The sun had gone down and the seaplane base lay on the islet of Nisida, 300 yards off the mainland. He remembered that telegraph wires spanned this short stretch, but in the fast-fading summer light, he nearly flew into them. In fact, Cobham was pretty tired after his long months of preparation following right on top of his Capetown epic. He did manage to get across to Athens, however, where he stayed to rest for a day on doctor's advice.

Cobham was still not fully fit, yet went on to the rigid timetable he had set: Athens, 3 July; Alexandretta, 4 July; Baghdad, 5 July.

After they had crossed over from the Tigris to the Euphrates and gone about 150 miles, they ran into a thickening sandstorm. Bringing the seaplane to just a few feet above the river, Cobham hammered on through a blinding bank of choking sand. Finally it got too dense to see safely at the speed he had to travel, and he came down on the river beside a local police hut, beaching the plane on the mudbank nearby. This was the asset of the seaplane; he could set it down anywhere in the world, provided there was water. After a rest during the remaining hours of the morning, they took off to try and reach Basra that day.

After half an hour they met another sandstorm near the vast swamplands above Basra. Cobham did not feel too happy about flying over the murky mixture of mud and sand that merged into one and almost obliterated any sense of horizon. The storm was just as bad as cloying fog. All he could do was to screw up his eyes and strain to see the rushes and weeds growing from the swampy brown lake below.

After passing the town of Suke Shuyuk, Cobham took care to keep to the rather wavy shores of the lake-swamp, marked by the ghostly rushes rising out of it. He was at about fifty feet on a zig-zag course.

Slowly the swamp seemed to come to an end and a sandy coast appeared. Cobham thought they were over the worst of the storm and would soon be at Basra.

Suddenly he heard and felt a fierce explosion. It seemed to come from the cabin.

He shouted through the connecting window, 'What on earth has happened? Are we on fire?'

The only explanation Cobham could think of was that one of their rocket pistol cartridges had gone off, which would have set the plane alight.

Elliott called back feebly, 'Petrol pipe's burst.'

Cobham could hardly hear him. As he was still flying low, he could not switch off the engine and just glide, so as to hear him more clearly. He scribbled a note on his writing pad, ripped the sheet off, and stuffed it through to Elliott.

After a pause, Elliott sent back a message that the petrol pipe from the reserve tank in the cabin to the supply tank on the top wing had burst a few inches from the point where it was joined to the cabin tank, and that he was hit in the arm very badly and was 'bleeding a pot of blood'.

Cobham snatched a glance and saw how terribly pale he looked. Should he land and try to give first aid – or fly on for Basra? It was an awful decision to be forced to make, but it was really made for him when he looked below. Once more all he could see were shallow swamp waters, dark, dank and dirty. The heat was terrific and Cobham knew that even if he did land he would run the risk of beaching the machine and not being able to take off again without Elliott's active help.

Against this, he thought, 'Elliott is bleeding and I might be able to stop it.'

But this was the short-term outlook. Elliott obviously needed a doctor and if Cobham got caught in the swamp and the dust storm that still eddied all around them, then they might never get out of that appalling place again. So he decided to try and make Basra at maximum speed.

In a temperature of 110 degrees Fahrenheit in the shade, Cobham raced the seaplane on at fifty feet. He reckoned they were about a hundred miles from the city, so hoped to get Elliott into medical hands within an hour. The oil got hotter as the plane shuddered full out at 125 miles an hour, still skimming the bank of the lake. At last, thirty miles from Basra, Cobham flew out of the severe storm and the sun broke through scorchingly.

A quarter of an hour later the De Havilland aeroplane was angling for the wide Tigris at Basra. But as Cobham came down over the river, he found that the open bank he needed for beaching the aeroplane did not seem to exist along the heavily built-up riverside. Finally he found a clearing, landed, and beached the aeroplane up on the mud.

The second that Cobham opened the cabin lid, he realized Elliott was in a bad way. The afternoon heat had become unbearable, and Elliott was having difficulty in breathing. Cobham got one or two reluctant locals to help him out of the cabin, but as the pilot had Elliott in his arms, and was struggling to step down from the lower wing on to the floats, Elliott whispered, 'Turn the oil off.'

Elliott feebly tried to push down the lever close by him.

Cobham gave him some brandy and began a long fight for help. No one would do anything at first, but eventually he did get Elliott away to hospital. Cobham remembered Elliott saying: 'I can't understand how the petrol pipe burst.'

Meanwhile, the pilot returned to the seaplane and supervised its towing down the river into the backwaters of the RAF inland water-transport dock, where they moored it.

That evening Cobham tried to explain to the commanding officer there how the accident had happened, but the engineers insisted that the petrol pipe could not have burst, or even if it had done, could not have caused so much havoc. Cobham heard that Elliott was holding his own. He went to bed that night still wondering about the whole strange episode. Next morning he learned the truth.

'Did you see any natives about when you were flying over the swamp?' the engineer-officer asked Cobham at breakfast.

'No – we couldn't see anything at all for the dust storm.'

Then the officer broke the devastating news that natives must have been there – and shot at the plane.

'A petrol pipe didn't do the damage,' he explained, 'it was a bullet. It entered the machine, pierced the petrol pipe, and hit Elliott.'

So that was it. The officer took Cobham to the seaplane to prove it.

Cobham worked out that they must have actually heard the gun being fired by an Arab. Then the bullet passed between the two floats, pierced the wall of the cabin, shot straight through the petrol pipe inside the cabin, and went on through Elliott's arm and into his side, passing both lobes of his left lung, and finally lodging itself under his right armpit.

Cobham was just about to go to bed that night when a telephone message reached him: 'Tell Cobham that his engineer Elliott had a sudden relapse and died at 11.15 tonight.'

Alan Cobham was utterly stunned and decided to give up the whole project. He had known Elliott for a long time and flown tens of thousands of miles in his company. It was as if he had lost part of himself.

The authorities investigated the crime among the Arabs in the area of the storm and later on an Arab actually confessed.

But before this, Cobham had been persuaded to continue with his plans and so chose an RAF Sergeant, A. H. Ward, to accompany him from Basra. Ward was a Cockney with a sense of humour and Cobham certainly needed cheering up after the tragedy that had marred this great venture.

They took off on 14 July and immediately ran into more trouble. While flying over desolate swamp near the head of the Persian Gulf, the engine started to die. The indicator showed a fall of two hundred revolutions. Cobham could not account for it, and fortunately it did not last long, or they might have been marooned in the damp deserted wastelands. That was the only moment the engine faltered during the 28,000-mile flight.

The next panic came an hour or two later when Cobham noticed that the petrol was getting perilously low. He doubted if he could reach Bushire on the fuel left in the tank.

Ward pumped and pumped but still there was no sign of more petrol getting through.

'What's wrong?' Cobham asked.

Then as things began to seem desperate, he suddenly remembered that there was a petrol cock at the bottom of the pump.

'Perhaps it was turned off while we were at Basra,' he called.

Ward nodded and in a second or two that minor crisis was over.

The next crisis occurred when they had to come down on the water at Bandar Abbas while a very heavy sea was running. Cobham managed it. Then as the seaplane was being towed out, one of its wing-tips and one of the tail-tips collided with a launch.

Two days later at Chahbar, they had just refuelled in readiness to push on to Karachi when Ward stepped on a float that was not there and fell into a choppy sea, but he was no worse for his ducking. Another thing that the engineer had to bear in mind when working on a seaplane in the open water was that unless he had a sheet underneath the engine, whatever he dropped he lost!

There were plenty of incidents to keep Cobham's mind from dwelling too much on the tragedy of Elliott. At Bahawalpur, deep inland on the Indus river, they were just taking off when Cobham shouted, 'Look! There's a chap hanging on to the floats!'

Cobham throttled down and shouted to Ward to get out of the cabin and make the man get off. The only way to force him away was for the engineer to tread on his hands until he had to let go. The alternative would have been worse for the Indian, to be dragged up into the air as he would have been in another moment or two. After he splashed into the water, Cobham saw some Indians rescuing him. Ward scrambled aboard again and they were airborne.

They flew on eastward: Delhi, Allahabad, Calcutta, Akyab, Rangoon, which they reached on 27 July. Although they still had some thousands of miles to go, the flight went pretty smoothly from there on, with almost an air of inevitable achievement. By the end of July they were at Singapore, and then they flew on via Muntok, Batavia, Sourabaya and Bima to Kupang, touching down there by 5 August.

The last lap to Port Darwin represented the longest jump over the ocean of the entire flight – 500 miles to the Australian coast. They started out on this long, lonely vigil from Timor to Port Darwin at a height of 100 feet and never rose above it all the way. Timor means fear. They hoped they would have no cause for it.

Cobham flew straight at the south-east trade winds. After about twenty minutes from Kupang they had left land behind and resigned themselves to some four hours' flying towards a horizon devoid of anything save the sea and sky.

Cobham calculated the shortest time this could take, and also the longest. If they had not sighted land when the latter had elapsed, he would have to fly due south in order not to miss the Australian continent. He did not want to career on east towards the Pacific with only a limited fuel supply aboard.

He had heard that visibility in that region might reach the phenomenal figure of 150 miles, and so when he had not sighted land after a reasonable time, he started to get slightly worried. He reasoned to himself, though, that he could not see so far that day, and so persevered on the same course for some miles.

Still no land appeared. This meant that either the headwind was holding them back more than they had realized, or they were drifting north of the island of Melville, and missing Australia altogether. Cobham thought the wind must be to blame. The shadow of the seaplane darkened the coloured coral reefs, but still there was no land. The engine ran perfectly. But time was ticking away ominously and Cobham was concerned by now. He worked out how much longer the fuel would last and then wondered how he would land on the rather rough water.

After what seemed an infinity of miles and hours, Cobham glimpsed a dim kink on the horizon.

'Look!Land!' he shouted through to Ward.

Half an hour later the seaplane sailed over a sandy beach, backed by red cliffs which were topped by a bushy jungle. It was Herd Bay, only five miles from the point they had aimed to hit. A hundred miles and one hour later, they saw the harbour of Port Darwin.

Four fantastic weeks followed, from 8 August to 4 September, as Cobham and Ward flew all around Australia to tumultuous welcomes at each stopping place, large or small: Katherine Station, Newcastle Waters, Brunette Downs, Camooweal, Cloncurry, Longreach, Charleville, Bourke, Sydney, Hay, Melbourne, Adelaide, Oodnadatta, Alice Springs, and back to Katherine Station and Darwin.

Just as Cobham had taken his time on the outward trip to Capetown and then hurried home, he wanted to try and do the same from Australia and arrive back on 1 October – a dash flight, as he called it. He planned to make two jumps a day, averaging 700-1,000 miles. In Australia Cobham and Ward had been joined by a third crew member, C. Capel, for the return journey.

They left their twin float trails in the waters of Port Darwin on 4 September, and reached Penang four days later. They left Penang, and suddenly felt the full impact of the monsoon lashing across the Indian Ocean.

The rains slashed at the seaplane, as Cobham desperately dodged the worst deluges for fifty miles and more. Finally he found himself surrounded by storm-centres and forced to go through them. On the outward leg, he had been able to see sixty miles in this area; now he was lucky to see sixty yards.

The faithful plane flew on, buried in the banks of the storm. By now Cobham could scarcely see out at all, and he suddenly sensed a dark mass dead ahead of him. He did a steep vertical bank, just in time to avoid hitting a gaunt rock rising five hundred feet out of the ocean.

Later on he made an emergency landing on an uninhabited island forty miles from the mainland. They eventually got away from the island, only to find that the wooden propeller had been damaged by the downpour, so Cobham had to put down at Tanoon.

Getting as far as Victoria Point, they tried to forge on to make up time, but conditions were against them. The further north they flew the more the monsoon worsened. The force of the rain blinded Cobham, so that to steer on would invite disaster. The cascading rain wiped out all vision through his goggles, so he had to shelter behind his screen and just look out sideways, without getting the full fury of the rain in his face.

They turned back and were pleased to see Victoria Point again. They had to stay there, itching to be airborne, for four days while up to five inches of rain fell daily. They even had to bore holes in the floor of the plane to let the water out.

They saw the back of Victoria Point on 14 September and from then on made steady progress to Karachi. Then they really turned on the pressure, even through the oven-hot Persian Gulf, when the oil emerging from the engine registered 76 degrees centigrade.

The flight from Karachi to home took a mere week. When it was nearly over, Cobham was hurrying to keep to his schedule of reaching London on 1 October. During the flight over France to Paris he went down to an altitude of only a hundred feet over hill and dale to avoid the strong winds. He knew too well that if the engine failed or a fault occurred just then, they would have no time to find water and it would mean the end of a marvellous plane and perhaps themselves.

At Paris, Cobham collected a telegram asking him to land on the Thames in front of the Houses of Parliament at two o'clock the following afternoon, which he did. As a result of this fantastic flight of 28,000 miles in three months, including the return trip in under a month, he was given a knighthood. No one would deny that he deserved the honour.

8

BYRD AND BENNETT

to the North and South Poles

ADMIRAL R. E. Byrd was the conqueror of both the North and South Poles by air. These flights were fraught with hazards as bad as any in aviation history.

The attempt on the North Pole was made on 9 May 1926, and Byrd was actually only the navigator on this flight. The pilot was was another well-known American, Floyd Bennett.

In fact the air conquest of the North Pole developed into a dramatic race between the Americans in an aeroplane and the explorer Amundsen in an airship. The latter had been designed by the notable Italian flier, General Umberto Nobile, and was called *Norge* in honour of Amundsen's country, Norway.

The Americans' machine was a Fokker Type F.VII monoplane, powered by three 230-horsepower Wright engines. They had called it *Josephine Ford.* There could be no doubt that it was a Fokker for the word was painted practically all over the body and wings.

Both parties found themselves at Spitzbergen at the same time in early May, and it was a toss-up which type of flying craft would be ready for the honour of the first shot at the polar flight. The Norwegians and Italians had the harder job of preparing their cumbersome craft, but they laboured right round the clock to try and win the honour for the airship.

Meanwhile Byrd and Floyd Bennett had their own troubles. The Americans had to try and perfect special skis instead of wheels, so that they would stand a better chance of landing safely on ice, if necessary. Wheels would be useless at the polar ice-cap. But every time they tried the vital functions of take-off or landing they had accidents to the skis on either leaving or returning to the ground. Each pair that snapped or otherwise came to grief under the strain shortened not only their supply of skis but also the time available. They could see that the rival group were putting the last touches to their plans for the flight.

The night of May 8/9 proved decisive, for Byrd and Floyd Bennett knew the airship would be ready by next day. It was midnight. They fitted the final pair of skis.

'Let's have a shot at it,' Byrd suggested.

'Okay – I'm game,' came the pilot's response.

So at about 1 am in the Arctic twilight world, the *Josephine Ford* slid forward over the Spitzbergen ice and was airborne. Their rivals were left gaping and could only wait to see if the aeroplane was successful; the machine carried fuel for some twenty hours so they would know by the end of that day one way or the other.

The big 65-feet-span aircraft headed north. In this Arctic realm of no night, the twilight broadened and brightened into day. The two fliers sat muffled up like polar bears as the plane passed over the strange, silent, white world below. They had already gained valuable knowledge of Arctic conditions while flying over Greenland, and it was because of these flights that Byrd had developed some special navigational aids for the region: a bubble sextant and a sun compass.

Now the little huts of Spitzbergen lay far behind, together with the extra signs of civilization such as an odd ship or two, the airship's mooring mast, and wireless masts. Now they were well on their way along the 800-mile line to the Pole, which meant a round trip of 1,600 miles over the pale plains and uplands. The whole landscape seemed to merge into a single flat, white canvas – virgin, isolated, icy.

Byrd took some readings with his special instruments and announced: 'Should be nearing it now.'

Then they discovered in dismay that there was a leak from one of the oil-tanks. As so often happened, the fault at once presented them with a choice: to fly on and hope for the best or to try and land and see if they could plug up the leak. Either way they could be wrong, and to be wrong could so easily mean death.

Byrd bore in mind the trouble they had had so recently which resulted in those pairs of smashed skis. He did not want to risk that again, for they might never be able to get off the ice. But the alternative was not promising either. The leak would not improve by itself and would probably deteriorate, resulting in a forced landing. Whichever option they chose gave the prospect of possible disaster.

'Think we'd better stick it out,' Byrd said. 'No point in landing unless we have to.'

They kept going and the oil-leak dripped away, no better, no worse. Around 9 am, eight hours out from Spitzbergen, Byrd double-checked his measurements and reported, 'We're over it.'

While Floyd Bennett wheeled the plane in a steep circle, as if to encompass the hallowed area, Byrd threw out an American flag they had brought for the occasion, to signify the success of this first flight over the North Pole. They did not hang about after that, as they knew that although they had achieved their aim, they were only half way to the total distance which must be flown before regaining the safety of Spitzbergen.

Trouble met them as they headed south. As the aeroplane was tilting at a rather sharp angle, the bubble sextant slid off Byrd's chart-table to the floor of the cabin. It was no use to him or anyone after that, but nor was a normal compass so close to the Magnetic North Pole.

The resulting panic was merely momentary, though, for Byrd had his special sun compass, which served as a good substitute. As Floyd Bennett concentrated on the flying, Byrd exerted all his prowess in striving to see that they hit Spitzbergen.

They had been flying non-stop for fifteen and a half hours when the Fokker finally topped one of the white hills around Spitzbergen, a dark dot in a clear sky to the watchers. As it came nearer they could see that its sleekly tapered wings flew dead parallel to the icefield.

Amundsen bore no grudge against Byrd and Floyd Bennett, shaking them sincerely by the hand soon after the plane had landed.

Despite having been beaten, Amundsen went ahead with the *Norge* flight, and instead of making a there-and-back trip like that of Byrd and Floyd Bennett, he determined to become the first to fly right over the roof of the world.

So on 11 May, the Norge started their sensational flight from Spitzbergen to Alaska. They had a crew of sixteen and the airship was actually under the command of the Italian General Nobile.

The course to the North Pole was uneventful, but from there on they were over entirely unexplored territory. Then two things happened. Firstly their radio transmitter failed, so that the world had no news of them. Secondly, the great dirigible blundered into a thick bank of freezing Arctic fog, which persisted for ten hours. Soon after they ran into this it started settling on the airship as moisture; the moisture froze into ice on the metal portions of the craft; and it was then sucked into the propellers of the airship in the form of jagged icicles. The revolving propellers spewed these out viciously against the lower part of the actual envelope of the airship. The icicles ripped rent after rent in the vital fabric and gas started to seep out.

The Norwegian and Italian crew had to put patches on as many of these tears as they could. The work was appallingly hard in freezing foggy weather but they managed it, and so the precious gas was conserved and the airship kept aloft.

Right on across the unexplored wastes they floated, all the time nearing Point Barrow, Alaska, their goal. The Beaufort Sea, the loneliest water in the world, drifted away beneath the cigar-shaped craft and eventually they reached Alaska after nearly three days and nights in the *Norge*, which had covered 3,500 miles.

Although airships are really outside the scope of this story, any account of the exploration of the Arctic and Antarctic from the air must include the sequel to the voyage of the *Norge*.

General Nobile wanted to extend Arctic exploration by airship by building a new one that would enable scientists actually to land on the polar regions to collect data and pictures. He interested his government in the idea, and the prompt result was the *Italia* airship, faster and more powerful than the *Norge*.

By the time it was built and had made the long journey from Milan to Spitzbergen, two years had elapsed. The *Italia* reached King's Bay, Spitzbergen, and took on 1,500 gallons of petrol for its three engines, which gave it a speed of 70 miles an hour. One engine was situated towards the rear and the other two in wing cars. Each of the three cars could be reached via the massive keel of the airship. The total volume of the *Italia* was 653,000 cubic feet.

Aboard the 'semi-rigid' now were a Swedish scientist, Professor Malmgrem, and three Italian naval officers, Viglieri, Mariano and Zappi.

They hoped to land at the North Pole, as well as fly over it, but as it turned out, a weird whining wind was sweeping the snow into ominous drifts when they arrived, and so they dare not risk a landing. They dropped a wooden cross down, given to them by the Pope, and with it fluttered the flag of Italy. Their radio sent messages far across to the Mediterranean. They even played music on a gramophone.

That was the high point of the voyage, and after it came tragedy.

Fog suddenly shrouded the airship, the same sort of freezing fog that had threatened the *Norge*. It was also snowing. The fog froze into ice, and the ice stuck to the envelope. The airship got heavier; its controls slower and more sluggish.

They used all its engines throughout that night, but the wind worsened and the ice remained. With morning came only the dim sight of solid ice-packs below and solid fog above.

Everyone was tired but at least the airship was still moving. Then the elevators jammed.

The coxswain could not turn the wheel.

They stopped all engines while repairs were being rushed, and the vast volume of gas floated the airship up over the uncharted Arctic snowscape to 3,000 feet.

With the controls working again, they started engines and nosed down through the layer of filmy fog to 1,000 feet. All appeared to be well once more, but it wasn't. The ice on the envelope was forming more and more thickly, thrusting the airship downward at the stern under its deadening weight. It was too late to do anything but wait for the worst to happen.

General Nobile recalled it like this:

> 'The recollection of those last instants is very vivid in my memory. I had scarcely had time to reach the spot near the two rudders, between Malmgrem and Zappi, when I saw Malmgrem fling up the wheel, turning his startled eyes on me. Instinctively I grasped the helm, wondering if it were possible to guide the ship on to a snowfield and so lessen the shock. . . . Too late! . . . There was the pack, a few yards below, terribly uneven. The masses of ice grew larger, came nearer and nearer. . . . A moment later we crashed.
>
> 'There was a fearful impact. Something hit me on the head, then I was caught and crushed. Clearly, without any pain, I felt some of my limbs snap. Some object falling from a height knocked me down head foremost. Instinctively I shut my eyes, and with perfect lucidity and coolness formulated the thought: "It's all over!" I almost pronounced the words in my mind. It was 10.33 on May 25.'

He was in the control-car of the *Italia,* one of ten men there. The airship struck the ice-pack so hard that the control-car was wrenched off and left sprawling on the jagged ice.

Suddenly lightened by the loss of the car, and ten men, the *Italia* swooped up, shot higher into the air, faded into the fog, and was borne away on the wind higher and higher – with the remainder of the crew. It was a terrible nightmare. Nothing was ever heard of it again.

One of the ten men in the control-car was killed by the crash, and Nobile had been hurt. The radio had been damaged but they were able to repair it and send out distress signals.

They had camping gear and food with them, so things were not yet hopeless; but the ice around them might break up and with it might drift their pitiful little camp, and any hope of rescue. Before this could happen, three of the survivors decided to leave the rest and set out to try and get to the mainland. They were Malmgrem, Mariano and Zappi.

As soon as the radio signals had been received, rescue operations were set in motion. Amundsen himself arranged to be flown in a French seaplane to the area, piloted by Rene Guilbord. Tragedy struck again, for the two men never reached General Nobile and the rest, disappearing to their death at some remote point along the route.

There followed a long, long wait. Many days passed.

Eventually a large flying-boat penetrated to the marooned men, dropping supplies to them by parachute. The ice was beginning to break up badly then, though, so the flying-boat could not land and rescue them.

After that a Swedish flier landed beside them in a Fokker monoplane specially equipped with floats, but he could only take one man back. The others made Nobile go as he was injured. On the Swede's next trip, his plane became damaged and the rescuer himself was stranded with the others.

Meanwhile, all this time the ice-breaker *Krassin* had been cracking and crashing her way through towards them, complete with aeroplanes. First of all, the *Krassin* reached Mariano and Zappi, two of the three men who had tried to escape on foot.

They were exhausted, starving and all but dead. Malmgrem had collapsed earlier and insisted that they leave him and carry on alone. No one ever traced him.

At length, the *Krassin* forced her way through to the main party of survivors – six weeks after the loss of the *Italia*.

After that episode, Byrd's success over the South Pole the following year came almost as an anticlimax.

He was well equipped for this operation, with four pilots and four aeroplanes, housed at his base in the Bay of Whales, 400 miles from the Pole. Their preliminary survey flights were not devoid of incident and during one of these expeditions three of the men were trapped by an unexpected blizzard that blew their plane over, as if it were a toy model. The storm did not die down for over a week, and all that while the men had to exist flat out amid the scant shelter of the Antarctic mountain. Byrd eventually rescued them in a Fokker monoplane.

They loaded provisions for several months into the Fokker aeroplane, which was powered by three Ford engines. The group comprised Admiral Byrd; Balchen, the pilot; June, the telegraphist; and MacKinley, the photographer.

Everything went well. They flew up a huge ravine, to avoid attempting to top the highest peaks in the Queen Maud range, but they had underestimated the extreme concentration of down-draught rushing through the ravine. The aircraft wobbled and wavered. The wind worsened the nearer they got to the top of the ravine. Suddenly Byrd sensed that they could not make it, laden as they were.

So they started to pitch everything possible over the side: food, fuel, stores, the lot. With each load that went overboard, the aeroplane flew a little more easily and rose a few feet higher. They left a strange trail of assorted stores up that ravine, a slope never before seen by man. At last the plane shot the summit of the pass and they flew over a fantastic plateau.

They surmounted more mountains and all at once the four Americans were actually on top of the snowswept, ice-capped, gale-torn South Pole. The date was 29 November 1929. Byrd had made history twice by flying over both extremities of the earth.

9

CHARLES LINDBERGH

first Atlantic solo

LINDBERGH – the name seems to symbolize the spirit of the great pioneers. Charles Lindbergh was the first man to fly the Atlantic solo, but the Lindbergh story started years before that landmark in his adventurous aviation career.

Flying had always been in his blood. He flew first at the age of twenty in Nebraska and soon afterwards spent most of his time in that state doing 'barnstorming' flights, either giving the public paid flips or daredevil exhibitions. Wing-walking and other such stunts figured among the acts created to bring people to watch him.

It was 1923 when he raked up the money to buy his first aeroplane at a cost of precisely 500 dollars! After more barn-storming and more flights for the public at five dollars a time, the young Lindbergh reviewed his prospects and decided that he would get more aviation experience by becoming a flying cadet in the US Army Air Service. After this top-grade training, he became a second lieutenant in the US Air Corps Reserve of Officers.

But not before he had nearly been killed. Lindbergh had done quite a bit of parachuting during those earlier stunt days, but to qualify for membership of the Caterpillar Club, a man had to show that he had made an unpremeditated jump. In other words in an emergency. Lindbergh could prove that easily – several times over.

Twice during his actual service training, he had to bale out. On 6 March 1925, the aeroplanes of Lindbergh and another pilot collided during a mass dummy diving run on Kelley Field, Texas. Not only did the two aircraft collide, but they interlocked too. Inextricably jammed, they started spinning down with their young pilots. The two men wrestled with their respective controls for an instant or two, but soon saw it was hopeless.

Lindbergh jumped, and the other pilot followed his example, but realized that the planes were liable to hit him if he opened his parachute at once. As soon as Lindbergh felt fairly safe from the risk of the cascading planes, he pulled his ripcord and his free fall was jerked to a sudden stop.

Still tightly together, the spinning aeroplanes carved their way into the Texas soil far below, erupting into fire. Lindbergh watched them and thought how lucky he was not to have gone with them. Skilfully pulling on his parachute lines, he drifted down in a clear spot and saw the other pilot reach earth safely too. That was escape number one.

On 2 June 1925, came number two, while he was still with the Air Corps under training. Trying out a machine at about 2,000 feet, Lindbergh suddenly found the plane careering into a diabolical spin. He knew he had some seconds at that altitude, so did not abandon the aircraft at once. He did all he knew to level off, but nothing stopped the headlong spiral. Lindbergh finally gave up and found he had nearly left it too late to jump at all. He baled out when a mere 100 yards above the ground at St Louis, but his parachute opened in time and he was saved again.

Such incidents were all in the game during those days of early flying. A pilot risked his life all the time. It was a recognized part of the profession, in fact almost an attraction.

When Lindbergh left as a qualified service reserve officer, he went back to barnstorming, but with much more knowledge of the art of aviation. This was only temporary, for he heard he had been accepted as a pilot to fly the early air mails between St Louis and Chicago.

Lindbergh was overjoyed and proceeded to extend his experience still further, with practical routine flying month in, month out. Routine was hardly the word for it, though. Lindbergh had to overcome fog, ice, and everything else that the Mid-West weather presented. 'The mail must get through' might have been his slogan. There were at least two occasions when it nearly *didn't*.

He was flying through fog at night on 16 September 1926, when he got lost in mid-air above an invisible town and its aerodrome. Flares and searchlights on the ground could not pierce the foggy layer between the airfield and the flier, whose petrol gauge indicated nearly nil.

Lindbergh went on for a few minutes, hoping for a way round the fog, but there was none. In a last climb, with the final few pints of petrol, he nosed up above the black blanket. Then the fuel ran out. Since he could not hope to land the aeroplane through that total black-out down to ground level he baled out.

As he floated towards the fog, the aeroplane he had just left started spiralling down towards him and actually threatened to hit him. He pulled on his various shrouds and just managed to avoid the spinning aeroplane.

Through the fog-belt they both fell, and Lindbergh landed in Mid-Western corn. Then he went to look for his aircraft and found it with the mail safe. Lindbergh had used all his petrol to avoid any risk of fire when the plane crashed.

The winter was the worst time for flying the mail, as Lindbergh found out during that season in 1926–7. He did not get off to an encouraging start, either, for on 3 November 1926, he had to bale out for a fourth time to save his skin.

The temperature had dropped so low that after darkness Lindbergh had to fly through snow falling so thickly through the night sky that it was like three-dimensional flaky fabric. In addition it was misty.

First of all Lindbergh tried to land, but it was hopeless. He and his little mailplane were stormbound; he could not see the landing field and the petrol would not last long.

Lindbergh climbed to the colossal height of two and a half miles, through the wild weather. He jumped. The night air froze him as he fell amid the mixture of snow, sleet and rain. He hoped he would land somewhere soft. He didn't.

The parachute brought him faithfully through the storm-swept air, to come to earth on some barbed wire, but he was not hurt. Lindbergh spotted his tangled mass of a machine from the air the following day. He landed and retrieved the mail undamaged. Both times he had to bale out, the mail got through in the end.

It was during those days and nights on the St Louis–Chicago mail run that Lindbergh first really thought of flying the Atlantic solo. Once he had set his heart on it, he ploughed full ahead, managing to raise the necessary money quite easily from sponsors in St Louis. Hence the name of his monoplane, *Spirit of St Louis.*

He chose a monoplane in preference to a biplane for various reasons, including its ability to carry a greater load per square foot of surface at a higher speed. Lindbergh favoured a single-engine plane since although it was more liable to forced landings than one with three engines, it had much less head resistance and so a greater cruising range.

With the triple-engine aircraft, there was also three times the chance of motor failure – an interesting way of looking at it! If this happened in the early part of the flight, it would not cause a forced landing, but would mean dropping part of the fuel and returning to try again. Lindbergh talked quite blithely about 'landing' though this would probably mean death by drowning, if it occurred over the Atlantic.

He summed up his choice of engine like this:

> 'The reliability of the modern air-cooled radial engine is so great that the chances of an immediate forced landing due to motor failure with a single motor would, in my opinion, be more than counterbalanced by the longer cruising range and consequent ability to reach the objective in the face of unfavourable conditions.'

Lindbergh settled for a Ryan monoplane powered by a Wright Whirlwind engine developing 220 horsepower and having nine cylinders. He took immense trouble over every detail and reckoned that in the worst event the aeroplane would be kept afloat on the Atlantic for a while by air in the fuel tanks. That was his idea, anyway.

The makers took equal trouble over details for this flight, for not only was Lindbergh's attempt to be the first solo crossing of the Atlantic, and the first flight since Alcock and Brown, but it would span some 3,600 miles – nearly double their record-breaking effort eight years earlier. Lindbergh would have to do everything himself, including navigating, and in addition combat the chronic loneliness that must engulf anyone attempting such a desolate undertaking.

In fact, so unlikely did success seem that no one took the flight very seriously beforehand. Some even dubbed him the 'flying fool'. It was strange how a few days changed him from a foolhardy airmail pilot into a world-wide hero, but only typical of public opinion.

The clean-cut twenty-five-year-old American, Charles A. Lindbergh, spent long hours tuning up for the flight. Or rather, for the two flights. For the first 'warm-up' lap was to be across America from San Diego, California, to New York, via St Louis. The plane had been constructed in California by the Ryan firm.

Lindbergh left San Diego at 3.55 pm Pacific time on 10 May 1927, flew all night, and touched down next morning at St Louis after a flight of 1,600 miles. After a night's rest, he took off for New York and got there the same afternoon to establish a record for the flight from California to New York. The term 'flying fool' was hardly heard at all now. Even the idea of the totally-enclosed cabin and a periscope for all external vision did not raise so many objections. Lindbergh had already flown some 3,000 miles across America. There was no real reason why he should not do the same over the ocean – though it still seemed quite impossible.

At 7.50 am, 20 May, Roosevelt Field, New York, the real test began. Captain Lindbergh, as he was by then, was called in his hotel about 2 am. Outside, rain dripped off the window-sill of his room. But that did not deter him.

Lindbergh took two sandwiches and two bottles of water

with him, and crammed himself into the tiny pilot's 'cage' of the Ryan monoplane. *Spirit of St Louis* was painted just fore of the wing.

He grinned goodbye with a boyish wave to the thousands watching him leave. Then the plane gained speed swiftly down the artificial hill runway, but barely cleared the trees at the far end. Its total weight was 4,750 pounds and it carried 448 gallons of petrol. This would last Lindbergh for some time beyond the forty hours estimated by him for the flight.

Reports of his progress started to come in erratically. A wobbling aeroplane at 600 feet went over Rhode Island at 9.15 am. By 9.40 am he was over Bryantsville, Massachusetts, flying so low that people could read the markings NX-211 painted on the underside of the wing.

For several hundred miles, the aviator followed his route at an altitude of 100 feet or less over the sea, coming as low as ten feet at times. As *Spirit* sped on across the wide expanse of sea between Cape Cod, USA and Nova Scotia, Canada, Lindbergh ran into worse weather. And quite soon after leaving Nova Scotia, he met rain and even some snow. Taking the plane down to sea level and then up again, he looked in vain for better weather.

The aeroplane was sighted several times before leaving Nova Scotia. It reached Main-a-Dieu at 4 pm heading for Newfoundland.

As he flew further across that 300-mile stretch towards Newfoundland, the significance of his isolation could begin to be gauged. For not only was Lindbergh alone, but he had no radio, and no easy way of landing on the sea if he had to.

Furthermore, Lindbergh carried only elementary small-scale maps and a comparatively primitive outfit of instruments. His secret of successful navigation, however, later transpired to be that he had memorized all the definite details of his route – from New York to Newfoundland, and on across to Ireland and France.

The world was already excited about his progress, but as he

neared Newfoundland there was bound to be a long gap between there and the next news, from Ireland if all went well.

As he flew north-east over Newfoundland, the weather worsened hourly. He failed to spot the island's coastline at all, obscured as it so often was by the flier's enemy, fog. But he did glimpse the ground inland as he went over, since it was white with snow.

As the telegraph systems of the world whirred with comment and speculation Linbergh went on sitting blind in his little cubby-hole, occasionally looking into the periscope.

Just before darkness fell, while he flew low over the ocean, he had the shock of seeing floating ice jolting about beneath him. He thought he had better be careful and not fly so low, in case he flew slap into an iceberg. He did not want to end up the same way as the *Titanic.*

As fog wisped more thickly through the blue-black air, he became increasingly worried about the risk of hitting icebergs and he decided to climb. From just a few feet over the Atlantic, he made a long gradual ascent to the regions of 10,000 feet – only to find thick storm clouds instead of the fog. He tried aiming *Spirit* straight through a black bank of clouds, but sleet started to appear on the slender wings. He knew what that would mean if it continued so he changed course to try and find better weather, but had to settle for the fog throughout the rest of the night. Fortunately the darkness was fairly short due to the northerly latitude of his course. He spotted the lights of one or two ships far below.

Cramped in his cockpit, he began to find the flight monotonous, though he did admit that it kept him at full stretch, attending to the various jobs apart from the actual flying operation. By dawn Lindbergh had already finished practically all his water.

At daylight the weather took a turn for the better. The remainder of the sleet slipped off the wing. The improvement was short-lived, though, and more fog appeared in front of his periscope. The flier took the aeroplane down low to try and get below it, but couldn't. It was too risky flying so low, and he went higher again, relying only on his instruments for two more hours.

Lindbergh almost lost track of time as the aeroplane plodded on through the patches of fog, which, to his tired eyes, seemed to take on the forms of trees, coasts and other shapes so convincingly that he felt they must really exist.

Once the fog finally cleared Lindbergh decided to come down as low as he could, ranging from his minimum of 10 feet to a maximum of 200 feet. For the rest of that day *Spirit of St Louis* skimmed eastward for Ireland. As he himself said:

> 'During the day I saw a number of porpoises and a few birds, but no ships. The first indication of my approach to the European coast was a small fishing boat. Less than an hour later a rugged and semi-mountainous coastline appeared to the north-east. I had very little doubt it was the south-western end of Ireland. I located Valentia Island and Dingle Bay and then resumed my compass course towards Paris.
>
> 'In a little over two hours the coast of England appeared Then across the English Channel, striking France over Cherbourg. The sun went down shortly after passing Cherbourg, and soon beacons along the Paris–London airway became visible. I first saw the lights of Paris a little before 10 pm, and a few minutes later I was circling the Eiffel Tower at an altitude of about four thousand feet.
>
> 'The lights of Le Bourget were plainly visible, but appeared to be very close to Paris. I had understood that the field was further from the city, so continued out to the north-east into the country for four or five miles, to make sure that there was not another field farther out which might be Le Bourget. Then I returned and spiralled down closer to the lights. Presently I could make out long lines of hangars and the roads appeared to be jammed with cars. I flew low over the field once, then circled round into the wind and landed.'

Fantastic crowds of cars and people blocked Le Bourget

completely. Drivers clambered on top of their vehicles to get a better view of the night sky, swept by searchlights, shattered by rockets and storm shells. It was some minutes after 10 pm when they heard the first faint drone of *Spirit of St Louis.* Lindbergh announced his arrival by dropping a fuse and then after his circles of the airfield, he dipped in for a landing.

The plane taxied for a hundred yards or so. The crowd brushed aside the cordon of police and soldiers, smashed the fences and flimsy barricades, and acclaimed their hero: '*Vive* Lindbergh!'

Then they dragged him feet first from the plane. It was the most dangerous moment of the thirty-four hours since take-off.

10

CHARLES KINGSFORD-SMITH

conquering the Pacific

Southern Cross was the perfect name for the plane carrying Captain Charles Kingsford-Smith across the Pacific, from Oakland, California, to Brisbane, Australia. It was a Fokker monoplane fitted with three Wright Whirlwind 220-horsepower engines, and it had a wing span of 71 feet 8½ inches. The total petrol capacity amounted to a substantial 1,298 gallons.

With Kingsford-Smith were his co-pilot, Charles Ulm; Harry Lyon, navigator; and James Warner, radio officer. The trip was planned to be made in three stages: the first, Oakland to Wheeler Field, near Honolulu; the second to Suva, Fiji; and the third to Brisbane, in Kingsford-Smith's own country.

It was on 31 May 1928 that *Southern Cross* soared into the air from Oakland, San Francisco, for the first hop of 2,400 miles to Honolulu in the Hawaiian Islands. The aeroplane was well prepared for communication and navigation. There were three radio transmitters and two receivers, with a wireless direction finder as well. For navigation they carried two earth inductor compasses, a periodic compass and sextants.

This flight was destined to have its main climax at the end. In fact the first 2,400 miles, flown in 27 hours 27 minutes, was so smooth as to become quite monotonous to the crew of four.

The only excitement came almost at the completion of this initial hop, when *Southern Cross* radioed at 11.43 am on 1 June: 'We are heaving in sight of Oahu. It is going to be a race whether we make land before our fuel is exhausted.'

Before that time, Warner, the radio operator, had notified that his receiving apparatus had broken down and that he thought they were lost. Despite this gloomy guess, *Southern Cross* circled in to Wheeler Field at exactly 12.15 pm Pacific

Coast time. Five thousand people cheered as they landed and taxied down the field to the front of the reviewing stand. Kingsford-Smith emerged with his fellow Australian, Charles Ulm, and announced that their petrol supply had virtually dried up, all but a few gallons.

Southern Cross went on later to Barking Sand, Kauai Island, about a hundred miles from Honolulu, and then in the early morning of 3 June, took off on the second stage of the 7,300-mile epic, to Suva, Fiji.

The aeroplane was once more loaded to capacity with fuel, and Kingsford-Smith said that if they ran into dangerous winds they would attempt to land on one of the Phoenix Islands. If all went well, though, the Suva landing would be the next, on a 400-yard long cricket field with its fences removed. They were still keeping their smelling salts handy, in case any of them felt sleepy and needed reviving. So *Southern Cross* winged its lonely way for the next 3,200 miles.

The course for Honolulu had been roughly south-west. The route taken now veered nearer south. The weather was worse in the earlier hours of this giant stride of almost a day and a half. Rain beat against the three engines and rolled off the propellers, but the aeroplane throbbed on at an average of nearly 100 miles an hour. Sometimes it had to describe an arc to try and avoid the storms. Its name *Southern Cross* was lettered large on its body, but there was no one now to see it.

Radio messages gave a good idea of the changing conditions as they neared the equator. They had momentary engine troubles, storms, adverse winds.

'You can't stop us smiling,' they transmitted rather naïvely.

'We're dodging about to miss heavy clouds,' read another. 'A great game dodging these dark clouds.'

'Motor spitting.'

'All okay,' came a little later.

'One generator quit, only three hours out, no chance to charge battery, and headlights also; wireless transmitter, too, has failed, so now using auxiliary light.'

These minor mishaps passed, however, and Kingsford-Smith

was later able to signal his position as latitude 2 degrees south, longitude 170 degrees 33 minutes west, at an altitude of 1,400 feet. An hour and a half after that, they were over the Phoenix Islands, not needing to land.

Radio silence after that report began to cause apprehension in the small room at Le Perouse, Australia, where the messages were being received, but following a two-hour gap, *Southern Cross* transmitted: 'Doing fine. Been expecting to sight land, but none sighted yet. It is not so clear as one would like.'

This was followed by a message giving their position as latitude 5 degrees south, longitude 172 degrees 25 minutes west.

A full tropic moon shone out of a perfect Pacific night at Suva. Then another day dawned. *Southern Cross* radioed that they had been driven west of their course and were then north of Fiji with adverse winds and seven hours' petrol left. They had exceeded their average fuel consumption, but still hoped to make the Albert Park Sports Ground. They did and achieved what was then the longest non-stop ocean flight on record, 3,138 miles in a time of 34½ hours.

As soon as it became known that *Southern Cross* was about to take off from Naselai Beach, Suva, on 8 June, Australia became agog with excitement. The fliers had covered the first leg of 2,400 miles and the second of 3,200, and now they were poised for the third of 1,700. Could they accomplish this 7,300-mile marathon?

The take-off was at 2.50 pm from Naselai Beach. Ten minutes later *Southern Cross* saluted Suva and then struck south-west for Brisbane. And the messages started to pour in as the Australian people grew more and more excited.

> 'Well, here we are, on our way again. Everything okay. In 19 or 20 hours we shall be in dear old Aussie again. The landing at Brisbane will be the culmination of ten months' hard work and the realization of our ambition to be the first to cross the Pacific. After arriving Brisbane will leave next day for Sydney. Can assure Brisbane public we shall return after couple weeks' rest. Cheerio.'

And another message:

'Smithy is at the controls with a sandwich in one hand. Ulm is working radio, while alert, loyal, efficient colleagues are a few feet behind us. When we started the flight we wondered whether Australia would want to kiss or kick us on arrival, in view of so much adverse criticism against our undertaking, but we've been so overwhelmed with kind congratulatory cablegrams and wireless reports that we feel our fellow-Australians now agree we were right in sticking it, completing our self-appointed task. We had a long but narrow strip of beach to take off from at Naselai, with fairly strong cross wind, but, comparatively speaking, only light load, 880 gallons petrol, 32 gallons oil.'

At 5.25 pm: 'Since leaving Naselai we have had trouble with one compass; nothing to worry about, cheerio, one generator bad.'

At 5.50 pm: 'Cheerio everybody. Won't be too long now. There's possibility of dirty weather tonight. Ulm has relieved Smithy, so may be better rested for tonight. We're happy as Larry up here, cooee!'

At 6.30 pm: 'Will spare batteries, only one generator.'

At 7.12: 'Did not wish to worry anyone before while going through a storm, but now it is almost passed will admit that the last two hours we have been battling with the worst weather of the whole flight.'

At 7.20 pm: 'Encountering exceptionally heavy rain. Bumps in air pockets, frequently dropping 400 feet. Smith and Ulm wet through. Cold, no gloves. As soon as the moon comes out and blind flying ends for the night we will eat and have a spot of our emergency whisky rations. Too cold to write more.'

They were not happy as Larry any longer. The distance to Brisbane still measured a thousand miles. And the storm still raged around them.

That compass was really right out of service, so they had to trust the magnetic steering compasses. The moon did not come out. Smithy had to try and get above the dangerous combination of wind and rain, but at six and then seven thousand feet things were no better, in fact the reverse.

The wind-lashed torrent had torn through the windscreens of *Southern Cross*, shattering the glass all over the cockpit. Water flowed inside. Kingsford-Smith was no longer resting. He coaxed a climb out of the monoplane, but the weather went right up beyond their ceiling, and the higher they went the colder they felt.

This was a night they would never forget. After the smooth flying for the first 6,000 miles, it seemed as if they might be beaten in the last straight.

An electrical storm sparked off all around them.

Fierce flashes of fork lightning exposed the night wastes of the ocean in grotesque light. *Southern Cross* was the only object in the sky. All the time they half-feared they might attract one of those fatal flashes, for the plane seemed to be flying right through the storm-centre. This was something they had not expected.

They were frightened that flooding might cause the magnetos to fail at any time. They saw blue flames sparking across the leads of the three engines, through the soaked insulation.

Kingsford-Smith felt his hands getting increasingly numb, while the rest of the crew became just as icy. The rain still streamed in through the smashed screens, directly under the wing.

Those air pockets were awful. There were sudden sheer falls of three or four hundred feet through the wildness of the night, the crew never knowing when they would come, or whether the plane would ever stop the descent once it did start.

At 11.16 pm they reported briefly: '700 miles to go.'

At 1 am: '600 miles.'

But between 6.15 pm and 3.20 am, nine nerve-racking hours, Ulm only made one entry in the log of *Southern Cross*. That was how bad it was.

The nightmare of fireworks and sickening drops passed.

At 6 am they estimated that they were 250 miles from

Brisbane, but it was not until 10.13 am that *Southern Cross* finally fought its way to Brisbane.

The career of Kingsford-Smith cannot be left there, though, for he had many more successes, and failures too, as one of the pioneers in those early days of world-wide flying.

After several successful Australasian flights during the rest of that year, Kingsford-Smith and Ulm set out for a flight to England on 31 March 1929, which was destined to fail. They had H. A. Litchfield and T. H. McWilliams with them.

Soon after leaving Richmond, Australia, they lost their long-wave aerial, so that they could not receive radio messages, although they could still send them. This turned out to be merely the first in a chain of mishaps, for as a direct result of it they did not hear warnings broadcast to them of a sudden storm in central Australia. They flew straight into it.

Somehow they survived the night, but the storm did not die at daylight. Kingsford-Smith intended to fly round Cape Londonderry to reach Wyndham. They tried to get their bearings by dropping a message to local natives in a habitation below, but due to a misunderstanding they were directed the wrong way for Wyndham.

Kingsford-Smith realized that the petrol was bound to run out before they could reach Wyndham, but there was little to be done but to try to plod on. An hour later the engine gave its first signs of fuel shortage and the pilot knew he must come down somewhere very soon. McWilliams went on sending messages giving their position as far as they knew it, and then he told Sydney that *Southern Cross* was making a forced landing.

Smithy looked doubtfully at the inhospitable landscape, lying wild and uncharted below them. He brought the aeroplane down as best he could, but the surface he had chosen was mud made wet by the recent storm.

At once the wheels sank into the squelching swamp and the aircraft was trapped. So were they, for dense jungle enclosed the mud-ridden ground, not to mention waterways alive with alligators.

It seemed silly, but there was nothing for them to do but wait. It would not have been a good idea to leave the area, for at least their radio signals were being received and search parties would be beginning to look for them from the air.

Then the next blow fell in the chapter of strange accidents, when they looked for their emergency rations – there weren't any. Their main nourishment would have to be the quantity of special baby food on board which was to be delivered to Wyndham. Their only other food was small amounts of coffee, brandy, glycerine and snails. Starvation looked a likely prospect, failing rescue. They found a source of water nearby, though it was not very wholesome.

Though the plane was stuck solid, they would explore the immediate zone and managed to kindle a fire on a nearby hillock. The rescuers would have to be quick to be in time.

In the overpowering, humid heat of the interior, the men began to weaken as early as the second day after landing. Two days later they were nearly too exhausted to keep the fire alight.

Meanwhile the rescuers were doing all they could to trace them. The stranded airmen had evidence of this. They saw three aeroplanes at various times, but for some reason the searchers did not spot either *Southern Cross* or the fire. This was the last straw, but worse was to follow before their luck changed.

Weaker each day, they could only wait while the ceaseless struggle to find them went on. Two of their flying friends, Keith Anderson and H. S. Hitchcock, had joined in the hunt – and vanished as well.

At long last, on 12 April, the thirteenth day after the start of the ill-starred flight, a De Havilland 66 Canberra biplane spied them, just in time. Circling overhead, the pilot made an accurate drop of food for them. For four days this went on, with daily drops of supplies, and they began to feel better. Then on 15 April a light aeroplane landed on the sun-baked mud. Two more planes brought petrol on the following day.

On 18 April *Southern Cross* itself was able to rise again from the hardened mud-swamp and carry on to Wyndham.

The tragic sequel to the story was that in trying to find them, Anderson and Hitchcock had themselves been forced down in a dry desert and perished.

The rest of the career of Kingsford-Smith was full of ups and downs, including a number of triumphant first flights. He pioneered England–Australia air mail flights when a colleague in his own airways company took off from Sydney with the Christmas mail for England. The aeroplane crashed five days later, but Kingsford-Smith at once flew to the scene of the crash, transferred the huge batch of mail to his machine and forged on for England.

The weather was cruel to him, but despite fog and snow he touched down at Croydon airport on 16 December, just in time for Christmas delivery. This flight in less than a fortnight and at a moment's notice was one of his major successes. Another came in 1933 when he flew from Lympne, Kent, to Australia in the remarkable record time of one week, or to be more precise 7 days, 4 hours, 43 minutes.

Then in the autumn of 1934, Sir Charles Kingsford-Smith, as he now was, and Captain P. G. Taylor tackled the reverse run across the Pacific, from Brisbane to Oakland, California. Flying a Lockheed, they accomplished this by 4 November in the short flying time of some forty hours.

Kingsford-Smith's final, fatal year was 1935. His first setback was when *Southern Cross* took off from Sydney at 12.30 on the night of 15 May to fly to New Zealand with special Jubilee mail and other cargo.

About 7 am the next morning, in the middle of the Tasman Sea, a fragment of exhaust pipe snapped off from the middle engine, and shattered a blade of the right-hand propeller. Smithy took immediate action to avoid disaster and then wheeled round to head home for Australia on two engines.

Southern Cross had to face a fierce headwind and the left-hand engine started to overheat. Captain Taylor crawled out of the cabin and along the wing with a thermos flask, removed oil from the right-hand engine and took it over to damp down the left-hand one.

He did this a total of six separate times.

He succeeded. They had to ditch fuel, freight, a large proportion of the precious mail, but they limped back to Sydney alive.

Kingsford-Smith's last flight of all began on 6 November, when he was trying to beat the record for the flight from England to Australia. In a Lockheed Altair two-seater, with Pethybridge, he got as far as India – and then they vanished over the Bay of Bengal.

11

AMY JOHNSON

solo to Australia

THE first and most famous long-distance flight by Amy Johnson was her solo trip to Australia in May 1930. She had taken up flying in her spare time and had later given up her office job. She then raked up the £600 for a second-hand Gipsy Moth light plane and calmly announced that she was going to try and fly from England to Australia and beat the existing solo record for the flight, Hinkler's fifteen and a half days.

At the age of twenty-two Amy was not only a qualified pilot but also the only woman to hold an Air Ministry licence as a ground engineer.

She took off from Stag Lane aerodrome, near Hendon, where she had first been attracted to aeroplanes when passing in a bus! Five light aircraft of her club escorted her across to Croydon, the official starting point for all record attempts, and from there she ascended at 7.45 am on 5 May. Normally a two-seater, the Gipsy Moth had had its second seat removed to make room for extra fuel tanks. Amy would need all the petrol she could carry on her 13,000-mile flight.

She had christened the plane *Jason's Quest*. She left the Surrey suburbs behind with fuel for a range of over a thousand miles. Amy's first scheduled stop was the Aspern aerodrome, near Vienna, 780 miles away. She got there by 5.15 pm the same day, after nine and a half hours in the air.

The flight was not yet attracting headline attention in the press; there was just a brief note that she reached San Stefano aerodrome, Constantinople, at exactly the same time, 5.15 pm, the following afternoon. So far the flight had been without incident, but then things got hectic for a day or two.

Crossing the Taurus Mountains on 7 May, at a height of

8,000 feet, Amy found she could not get as much altitude as she wanted, so she clung to the railway line through the narrow pass amid very poor visibility.

From Aleppo to Baghdad on 8 May she had a nerve-racking spell. Midway between these two remote places, the little Gipsy Moth was tossed about in the grip of a gale. It might almost have been a shuttlecock: it certainly dropped like one, from 7,000 feet to 300 feet in ten minutes. Sand swirled all around her and she decided to land.

She brought *Jason's Quest* down in the face of a 50-mile-an-hour gale, but it would not stay still once it was actually on the desert. Amy lugged out her tool boxes and luggage and with sand-dust stinging into her eyes, she shoved it against the wheels of the machine as chocks to stop it being blown away altogether and leaving her stranded an unknown distance from Baghdad.

She struggled to cover the exposed engine, too, so that sand should not penetrate and damage it. Alone there in the desert, she heard the eerie barking of dogs but saw nothing.

Two hours later the weather relented a little and she was able to take off again. She spotted the Tigris, the surface choppy from the storm and was able to follow its course. Atmospheric conditions then worsened so much that she missed Baghdad completely. The junction of the Tigris and Diala told her that she had overshot her target. Amy swung the little aeroplane around and later set it down neatly on the Imperial Airways aerodrome. When she bobbed out of the open cockpit, she was covered from hair to shoes with a thick coating of sand, evidence of the ordeal she had endured.

Amy flew on for Bandar Abbas, where landing was less easy, and she felt glad to get away from the fantastic heat of the Persian Gulf. Another lap of 720 miles took her to Karachi, where she arrived in six days, ahead of Hinkler's time for the flight to this first point of India. Now people started to take increasing interest in the young Englishwoman's exploit. She spent that night at Government House.

The public now learned that the Gipsy Moth had already

done service on long flights well before Amy bought it. Captain Hope had used it to follow the Prince of Wales on his recent African tour, and it had since been flown by an air taxi concern. Its mileage had exceeded 35,000 before Amy left home in it.

She found herself short of petrol on the way to Allahabad, so spent the night at Jhansi. Landing there in the strange Indian dusk, she struck a post with a wing of the plane, which was damaged. An Indian carpenter managed to repair it promptly, however, and she proceeded to Allahabad and thence to Calcutta, which she reached on 12 May. She was still well ahead of Hinkler's time.

She made a 7 am start on 13 May from the Dum Dum aerodrome, Calcutta, for Rangoon. Before leaving Amy made the modest statement, 'This is just an ordinary flight, except that it is longer. Every woman will be doing this in five years' time.'

She added that as she had to overhaul the machine herself at various stages of the journey, she had only managed to average three hours' sleep a day since leaving London.

She seemed to be unlucky in her landings. When she was about nine miles north of Rangoon, she mistook the playing fields at Insein for the Rangoon racecourse, her appointed field. This only happened because of bad visibility and forced her to fly at 200 feet or less, brushing the treetops with the exhaust of the plane.

She made a perfect landing at Insein, but as her machine taxied across the playing fields it ran into a concealed ditch which slightly damaged the wings, wheels and propeller. Amy did not intend to let a little mishap like that delay her, but she did, as it turned out, have to wait an extra day or two there, for the weather was misty and windy and the wings and fuselage needed repairs after the jolt in the ditch.

Amid blinding monsoon rain, 16 May dawned, yet Amy had set her heart on taking off for Bangkok – so she did. She left Rangoon behind at 10.45 am, knowing that the delay had cost her all reasonable chance of breaking the record. There was still plenty to fly for, however, apart from the mere fact of finishing what she had started.

Amy set the nose of *Jason's Quest* south-east heading for Moulmein, but the vital pass through the mountains eluded her, being hidden by the dense monsoon clouds. Try as she might, she could not find a passage, so she dragged the laden aeroplane up to 9,000 feet to cross the mountains blind. With this kind of manoeuvre, the risk always existed of miscalculating either the height of the mountains or of the aeroplane. Either would have been instantaneously fatal. Amy just kept to her course through the swirling skyscape. Looking out, it was as if the aeroplane were a toy lost in a grey world of fluffy, floating cloud-base.

Eventually she staggered out into clear weather again, only to have the shock of realizing she was still on the Burmese side of the mountains. She made another, more frantic, shot at finding the Moulmein, still totally invisible, and this time she did come out on the Siamese side of the heights.

She was not out of trouble yet. For three more weary hours, fighting sleep all the time, she could pick up no landmarks at all, roaming on and on through the layers of storm clouds. Then she espied a railway junction and all was well.

For six hours she had been lost in a fabulous world of monsoon-racked mountain tops between Burma and Siam.

Amy touched down amid the quaint curving temples of Bangkok at 5.45 pm on 16 May, dead beat but still game.

Two days later she got to Singapore. Between Bangkok and Singora Amy had run into bad weather once more, and for the whole of the hop from Singora to Singapore, she had to fly at under 1,000 feet. Two seaplanes met her seventy miles out, escorting her to the aerodrome at Selegar on Singapore Island. Amy landed at 1.50 pm grinning at the cheers of the Malays, British, Chinese, and other nationalities there.

As the 13,000 miles gradually reeled off, the flight assumed proportionately greater interest to the world at large. To succeed would be wonderful; to fail now, awful. Amy Johnson. The name was on everyone's lips in every tongue.

She left Singapore on the morning of 19 May bound for Sourabaya. But she was destined to have a dreadful time over the desolate Java Sea south of Singapore.

The rain literally hosed her off her course towards Sumatra and she had to fly over the miles of thick jungle which covered that huge wild island. Towards Banka, a large island off the mainland of Sumatra, she veered out to try and find better conditions over the sea.

Torrents of tropic rainfall soaked her and swept her down to mere feet off the whipped-up sea surface. She only had to lose a little more height and hit that water once for it to be all over.

'I really thought this was the end,' she said later, 'but somehow I reached the coast of Java and followed the coastline till I ran out of fuel.'

It was just one thing after another.

The heavy weather resulted in her using too much petrol – and too much time. She was virtually out of fuel and *Jason's Quest* was still almost an hour from Sourabaya.

Amy scoured below for somewhere to land, failed to notice a military emergency landing ground in the neighbourhood, but did see a small clearing. It was that or nothing.

Reappearing over Tegal, which she had passed a little while earlier, she circled several times before deciding that this piece of ground recently cleared for building a new house for the local sugar estate manager, was in fact long enough.

Amy deposited the aeroplane and herself on its short strip of 250 yards – the usual minimum was 300 yards – and accepted an invitation to stay the night there in the middle of Java as casually as if she were back at home with her parents in Hull.

Amy discovered the wings of her aeroplane were perforated in five places.

Bamboo poles supporting young fruit trees had caused these rents as she had come in to land. The holes were only small but the slightest flaws could be fatal. So the holes were patched up temporarily with sticking plaster!

Next morning, with a small supply of fuel to carry her on, she prepared to take off for Semarang. It was going to be touch and go getting airborne, however, from the wet, short clearing, so the estate manager kindly took part of her baggage to an emergency landing field five miles away, where she could pick it up en route.

Jason's Quest just succeeded in skimming the ricefields surrounding the Tjomal estate, and Amy was on her way again. The biplane reverberated on for Semarang and fresh fuel.

These delays were of course inevitable in an epic journey on such a scale. The miracle was that she had got that far at all in a Gipsy Moth costing £600!

At 11.30 am Amy left Semarang for Sourabaya. But before she did so, she said in a tired tone, 'I want so very much to rest.'

By the time she reached Sourabaya, the strain was starting to tell not only on her but on the faithful *Jason's Quest* as well. After a trial run in the afternoon shortly after getting there, she did not feel happy about the state of the machine and so organized repairs to the engine late that night.

By dawn it was running sweetly and at 6 am on 22 May she winged up from Sourabaya into a blue sky. She intended to fly for Atambua on the north-west coast of Timor Island, but bore in mind Bima on Sumbawa as a possible interim landing place.

The suspense by this stage was growing unbearable. Would she do it? Could the machine last out?

She had told spectators that the blue sky was a happy omen for her flight, and she was still waving to them several hundred feet overhead. Then she was gone, bound for Atambua, 750 miles off. It was a nasty span, mostly over the sea.

At 11.30 am she passed over Bima, but did not land. She was due to arrive at Atambua about 4.30 pm the same day.

The wireless station of Kupang on the island of Timor was kept open until 11 pm but then reported no news of her.

The whole world held its breath.

Surely she couldn't fail now?

Next day the news filtered through, 'Amy Johnson landed at the village of Halilulik, on the island of Timor, and went from there by motor car to Atambua, her original destination, which is about 12 miles away.'

The village had no telephone link with the outside world, which accounted for the delay in the news of her safety. She managed to send news of her arrival from Atambua in time to stop a naval aeroplane leaving Sourabaya in search of her.

It was not until later that she told how when she had landed on Timor she was met by 200 natives. They crowded round the aeroplane but she could not communicate with them. One of them took her hand and led her three miles through the jungle to a church, where she met the local parson. After that she was soon bumping over a twisting track to Atambua.

Then there remained the last and most hazardous hop of all. Amy Johnson had flown her *Jason's Quest* from Croydon to Timor. Ahead of her lay the 500 miles of sea separating that island from Australia. She had done over ninety per cent of the trip.

The 100-horsepower engine of the Gipsy Moth had brought her to the very edge of everlasting fame.

Amy took off from Timor, and set her course roughly south-east across the empty, liquid miles. She was utterly exhausted and cut off from the world.

Second after second passed, minute after minute, and hour after hour, with nothing below but water in all directions.

Amy flew on for Australia, across the sea of fear – Timor – as other fliers had done before and would do again. But until now no woman had ever done it alone.

It was at 3.50 pm on 24 May, Empire Day, that *Jason's Quest* formed a dark dot in the sky over Fanny Bay aerodrome, near Darwin.

Amy saw land and shouted for the sheer joy of it.

At 3.57 pm she was down. Cheers came in waves over the hot afternoon air; cameras clicked; she'd done it.

Dressed in khaki shorts, puttees, and a green sun helmet,

and looking sunburnt and tired, Amy said: 'Don't call me Miss Johnson. Just plain Johnnie will do. That's what my English friends call me.'

So Amy Johnson became the first woman to fly alone to Australia. Her time of twenty days was made in spite of two forced landings and an ignition fault corrected at Sourabaya.

As soon as her arrival was known in England, King George V sent the following message to the Governor-General of Australia: 'The Queen and I are thankful and delighted to know of Miss Johnson's safe arrival in Australia and heartily congratulate her upon her wonderful and courageous achievement: George R.I.'

12

JIM MOLLISON

the greatest solo ever made

THE greatest solo flight ever made, this is how Jim Mollison's epic east–west Atlantic crossing has been described more than once.

G-ABXY were the letters painted on the body of the small monoplane, which bore its name immediately beneath the cabin window: *The Heart's Content.* Jim Mollison's little silver streak of a plane was a De Havilland Puss Moth, measuring a mere 25 feet in length and with a wing span of $36\frac{3}{4}$ feet. Its other vital statistics were: petrol capacity 162 gallons; cruising speed 110 miles an hour; powered by a 102-horsepower Gipsy engine.

When fully loaded with petrol, no suitable airfield could be found in England, so early in August 1932, Jim Mollison took his little Puss Moth out of its hangar at the Stag Lane aerodrome and with only a nominal volume of fuel, flew it across to Portmarnock Strand, Co. Dublin, where there were several miles of flat beach at his disposal for a take-off with a full fuel load.

The weather was bad for early August and he had to endure day after day of waiting for it to break. The full moon came and went in the middle of the month. Then a heat-wave hit London and the Irish and Atlantic weather got better, too. The temperature leaped to 96 degrees in London.

Mollison had married Amy Johnson shortly before the flight and she was there to see him take off at about 11.30 am on 18 August. He had been waiting all this while for the surface winds over the Atlantic to lessen below 20 miles an hour, for these meant headwinds of twice that strength at the altitude he would be keeping.

He carried no wireless and he reckoned his fuel would last for a maximum of thirty-three hours. He felt sure this would be long enough as he expected to reach Harbour Grace, Newfoundland, within twenty-four hours. But he had to be careful about those headwinds, which could cut the actual aeroplane speed fatally.

His two compasses would not be affected by the magnetic disturbance off the Newfoundland coast if he maintained a minimum height of 1,000 feet. His total rations comprised some chicken sandwiches, a flask of coffee and some sticks of barley sugar for energy.

In fact the flight was intended to form the outward leg of a two-way ocean crossing within three days, later abandoned.

Navigation seemed to be the main difficulty and danger for Mollison. A pilot could not afford to make much of an error or be forced far off course by elements such as fog, cloud or wind. Apart from direction, there was always the question of pilot fatigue. To keep awake and alert for twenty-four hours is never easy and to fall asleep would be the surest way to die.

The Gipsy III engine was being tested just as severely as Mollison and if it succeeded it would prove itself fit for anything, since the east–west transatlantic crossing was much more difficult due to the necessity of flying against the prevailing south-westerly winds.

If Mollison made it, he would be the first person to fly solo westward across the Atlantic.

Loaded with its 162 gallons, the slender silver monoplane pressed its wheels hard into the mixture of sand and shingle as it gathered speed and finally rose from Portmarnock Strand.

The crowd stood dotted across the shore, cars intermingled with the people. A lot of spectators ran after the plane, as if helping it on its way. Then it shrank to a speck, vanishing over the sandhills in the direction of Malahide and the west. Soon after midday Mollison was reported over Galway at 1,000 feet. Then there was silence.

He was in an aeroplane of lower power than any which had previously attempted the ocean flight. Mollison's plan was to fly close to the water in daylight and rise only to a moderate height at night. He wanted to avoid blind flying if he could, although he had the usual instruments of those days.

Keeping low by day, his outlook was scarcely any more than that of a ship's captain from the bridge. With all these restrictions, it was clear that the weather could play considerable tricks with the course he hoped to keep. Just a small angle of error or drift through headwinds could fan out and amount to a substantial mileage at the end of the ocean journey of 2,000 miles. In fact Mollison's drift amounted to less than ten per cent.

The last ship to report seeing Mollison overhead was some 800 miles out from the west coast of Ireland. Then darkness spread rapidly behind him, chasing him from the west, and he was entirely on his own.

The drift south from his correct course had begun, caused by variations in the strength and direction of the wind. The great danger as always remained that of being swept too far south into the wider wastes of the central Atlantic and then for the plane to run out of petrol before sighting land. Even if the conditions were perfect, having to navigate as well as fly was bound to make the trip an exacting business.

Mollison left the ship behind and headed into the afterglow of sunset. This was the worst stretch, night-time in mid-ocean.

With darkness his horizon clouded so that after the first twelve hours of comparative comfort, things really became trying. As he passed the half-way mark and flew on, mist cast a net over the night. At daylight the situation had not improved and Mollison started to fear that he was not making the 100-miles-an-hour average speed he had reckoned as reasonable. He knew he must be nearing Newfoundland, but nothing more. His original aim had been to fly from Dublin to New York via Newfoundland, yet here he was in the dim morning of 19 August and for all he knew he might be over the St Lawrence or anywhere.

Now the infamous Newfoundland fog-belt really lived up – or down – to its name, blanking out the entire horizon, the seascape and skyscape. Not only seamen but aviators, too, had learned to hate this fearful fog, especially since it cloaked the ocean in the very area where a flier could begin to congratulate himself that he was near to attaining the goal of his flight.

Mollison groped, peered, swore, through the wispy vapours that by their very nature sent a shiver of unreality through anyone in their midst. He scanned the sealine for a break in the fog or the shape of land. He found neither. He reckoned he must have just about reached the coast but he simply dared not descend to water level without being sure. Firstly, he might make a mistake and sink to a watery grave. Secondly he knew enough about that craggy coastline to appreciate that unless he actually saw land he might easily strike on the frequent peaks pointing up from the island's shore. So he strained to see something, anything, and to avoid flying into a rock-face at nearly 100 miles an hour.

Finally he found Harbour Grace, Newfoundland, but decided not to refuel there as he had originally contemplated. Instead he circled the aerodrome, headed west once more, and gave the islanders the sight of the *Heart's Content* plunging into a thick fog over the Bay of Fundy. In fact, Mollison had nearly missed Newfoundland altogether by his southerly drift.

Next came that awkward sea stretch before he would be over Canadian soil and more or less safe. He made Halifax, Nova Scotia, all right, however, and had now been aloft and awake for approaching thirty hours. The mist was even veiling the land.

'How long can he keep it up?' everyone was asking.

On he flew over New Brunswick. He had used 152 gallons of petrol. He had ten gallons left, just about equivalent to a tank-full in a big car, and he was dog tired. The time was coming up to 12.45 pm New York daylight time, 5.45 pm British Summer Time. He had been up for 30 hours 15 minutes. That was long enough, he reckoned.

Jim Mollison thought he had insufficient fuel to reach New York, so he decided he might as well land wherever he saw a good spot. It came at Pennfield Ridge, New Brunswick, where the little Puss Moth fluttered down through the mist in a meadow.

At once he telephoned to St John, fifty miles away, where a message was relayed to his wife Amy, then awaiting news back in London.

This may have been the most outstanding of all Mollison's flights, but it was not the most eventful. He had several spills throughout his career, none worse than the joint trip by Jim and Amy Mollison across the Atlantic. But before this, Jim chalked up yet another amazing record, becoming the first person ever to fly from England across the South Atlantic to South America. The flight also marked the first solo success from east to west across the South Atlantic. Mollison used the same aeroplane as he had for his North Atlantic triumph.

He was still hankering after that non-stop flight from London to New York, having the foresight to realize that the air traffic of the future must flow between these two great cities.

De Havilland's built a big airliner for the venture, a Dragon twin-engine machine capable of seating ten passengers. For this flight, however, these seats were not included, cylindrical fuel tanks being installed instead to give the aircraft a petrol loading of 600 gallons and a range of 7,000 miles – unheard of for those days. Mollison had visions of longer journeys than merely the London–New York run, though for the time being this was the first object.

Jim and Amy christened their new plane the *Seafarer* and together they set off from Croydon in July 1933. The airliner had travelled two hundred yards along that famous early runway nestling amid the built-up environs, when suddenly its undercarriage collapsed – and that was the end of that.

Neither of the Mollisons had been hurt by the mishap and they felt impatient to try again. By August *Seafarer* was fit to fly but they preferred a sandy start rather than the Croydon field. Accordingly they conveyed the plane down to Pendine Sands, South Wales, which presented a seven-mile expanse of firm sand for the take-off. Jim remembered the successful ascent from Portmarnock and so felt happy that they would get away well this time, too.

Eventually the weather cleared enough to justify their departure. They had cut down the petrol to some three-quarters of the maximum possible, to lessen the aircraft's weight, but this 450-gallon capacity still allowed them ample for a trans-atlantic trip, or should have done.

Furlong after furlong *Seafarer* sped over the hard Welsh sands, while the cliffs rose in oppressive shadow to the land side. Half a mile passed and they were still on the beach. At last, after 1,000 yards, they got up and away. Then they nearly hit one of those rocky cliffs which loomed out of the sea-mist.

Anyone flying westward across the ocean was almost bound to encounter headwinds, and they were no exceptions. The wind slapped them the whole way over and made the forty hour ordeal even more exhausting. They had actually rigged up a simple bed in the airliner, but although there were two of them to share the work neither slept for an instant. They did take turns at the controls, but there was always something to be done even when not flying the aeroplane.

The climax came at the other end. Even thirty years later, the danger in flying is mainly on take-off and landing. At their first shot from Croydon, the take-off had failed. Now they were due for a disastrous landing just as they seemed assured of success.

For some forty hours, without a snatch of sleep, Jim and Amy had kept *Seafarer* on its route, determined to reach New York. But as they reached the United States and started flying over the New England countryside dozing below them, they suddenly knew that the few gallons of petrol left would not be enough. The headwinds had done their work.

They were only twenty minutes short of New York when they ran right out of petrol.

Bridgeport aerodrome lay beneath them and Jim started to coast and go into a glide for it. Lower and lower they sank, but it became clear to the group of onlookers that *Seafarer* would just miss the actual airfield. There was nothing Jim could do but put it down wherever he could.

The aeroplane struck a swamp adjoining the airfield, it swung over in two terrible somersaults and spreadeagled into a wreck.

Glass splinters ripped into Amy. Jim catapulted through the glass windscreen and pitched head-first into the slimy swamp. Extensively cut by flying fragments, he fell unconscious.

They were both hurried to hospital, where Jim Mollison needed a hundred stitches in his scattered wounds.

They had proved their point by all but twenty minutes. It was possible to fly from Britain to New York. Next time they would take more petrol.

13

AMELIA EARHART

first woman across the Atlantic

TALL, slender, softly-spoken and gentle-mannered, Amelia Earhart made several flights that rank among the finest flying achievements of all time.

She was the first woman ever to cross the Atlantic by air, but this was hardly one of her major deeds, since she merely sat in as a passenger helping with navigation on that memorable flight. It still had its interest, however, for the final few minutes.

Amelia had learned to fly several years earlier, though she did not take the controls on this trip, which was her first experience over the ocean.

She accompanied Commander Wilmer Stultz, the pilot, and Lew Gordon, acting as mechanic. They chose a Fokker seaplane with three 200-horsepower Wright Whirlwind engines. The machine was a monoplane with floats fitted.

They took off from Trepassey, Newfoundland, on 17 June 1928, and the *Friendship*, as the aeroplane was called, commenced the long flight. The details of the flight do not form part of this story. After hours of fog, the morning of 18 June brought clearer conditions.

Stultz saw a large vessel below on an approximately south-north route, which puzzled the three of them considerably.

'Where on earth are we?' Stultz asked.

Where, indeed, after nineteen and a half hours in the air?

Then minutes later another liner loomed over the horizon. They tried to contact her by radio without success. The mystery remained. Ships of this size did not steam on that course just off the west coast of Ireland, where the aviators reckoned they must be. But perhaps the very recent fog had literally clouded their calculations. After all they had not seen a thing until this last hour.

Then while they were still wondering where they were and what to do, they spied the dim outline of land on the easterly horizon ahead of them. They realized that it couldn't be Ireland.

In the long fog-belt, they had flown right over Ireland – the coastline threading in and out before their eyes was Wales.

Half an hour later the seaplane slowly scored twin ski trails in the waters of Carmarthen Bay and they had done it, but better than planned in fact.

Amelia Earhart had to wait nearly four more years before she could fulfil her real ambition: to fly the Atlantic alone and so become the first woman ever to do so. She would also be only the third person to accomplish it solo after Lindbergh and Hinkler.

Amelia looked a fragile figure as she climbed into her Lockheed Vega monoplane that evening of 20 May 1932, at Harbour Grace, Newfoundland. In reality her frailness concealed capability and courage of an extraordinary order. She was now the wife of an American publisher and so really Mrs Putnam, but the world will always remember her as Amelia Earhart – gallant, graceful, epitome of an era of flying forever gone.

Some flights seem destined to be smooth, others rough, and this was one of the latter. It was as if she had to be tested to the extremes of endurance.

Amelia revved up her 420-horsepower Wasp engine, opened the throttle wide, and almost before the crowd knew it, she was off, winging into the void at 7.30 pm on that mild May evening.

Three hours fled by and Amelia still felt fairly fresh. Then, about four hours out from Newfoundland, several things happened almost at once. Any one of them would have been enough to worry her, but together they seemed to spell destruction.

One section of the heavy exhaust manifold on the powerful

Wasp engine fractured slightly, began to leak, and let hot gases pass through. Tongues of flame licked momentarily over the cowling. By day Amelia would not have thought too much of this, but at midnight it was frightening, the blaze of the gases adding to the general fantasy of night over the ocean.

This alone was bad enough, but then she realized her petrol gauge must be leaking, and she could not remedy it. Fault number two.

The scarlet-painted Vega, its slim lines matching Amelia's own elegance, flew on and straight into foul weather. She began five long lonely hours of blind flying through the moonless night, with the aircraft leaving a trail of gas behind it. If Amelia shuddered, she would have been forgiven by anyone. Why was she here? She scarcely had a chance to think about that, which was just as well.

The thick, throttling clouds seemed to press in on the plane, uncanny and claustrophobic.

Then, to add to the flaming exhaust, the steady leak of fuel, and the five hours of flying blind in wind-whipped rain clouds – her altimeter failed.

'For the first time in ten years of flying,' she said.

So now was added to her existing troubles the hazard of not knowing how high she was flying, or whether she was level.

Flying blind without an altimeter was dangerous, she knew. She did not intend to run the risk of finding the plane diving into the stormy sea, the sea that might be any distance beneath her.

She had no choice but to climb right into the thick of those oppressive layers of clouds. She continued to climb until the tachometer – engine revolutions counter – froze from the drop in air temperature, and then at least she knew she could not possibly be near the sea. This was all very well, but the extreme chill did more than merely freeze the tachometer, it started to ice the wings of the Vega. This new development could also be fatal but she had to stay at this height.

The clouds were too thick for the aeroplane to penetrate them, so she did her best to pierce a path above them, not only for the simple benefit of finding and flying in clear weather, but to be able to check her course by the stars. The night was still upon her, seemingly lengthened far beyond its normal span.

She guessed that she was up to 12,000 feet. The ice worsened on her wings and the clouds still showed no signs of thinning. She thought of Hinkler, who had found himself in similar conditions and had done more or less the same as she, ploughed on through clouds so concentrated that they appeared to be solid.

Then lightning started shooting through the clouds. The leak in the exhaust manifold worsened, as the intense heat at the leaky joint began to burn away the metal. No longer did the burning flecks of gases flicker away into the night.

'I looked over at it,' said Amelia, 'and saw the flames coming out, and I wished afterwards I hadn't looked, because it worried me all night.'

As the flame-flecked plane proceeded, this trouble naturally grew worse. A piece of the manifold became detached and fell into the sea. Other parts of it began to work loose and serious vibration was set up, which could be felt throughout the structure of the whole machine.

The overheating of the manifold also started to affect the engine and it began running rough. She did not feel any easier knowing that petrol was leaking steadily into the cockpit from the faulty fuel gauge. She feared that petrol fumes would reach the manifold and be exploded by the flames pouring from the gap in it. Amelia had been in a variety of tight corners, but never one as bad as this. Fire was the flier's ultimate enemy.

Early that morning she flew over a ship which sounded its siren at her in greeting. She never saw another one. It did not occur to her to give up.

Now she had passed the half-way mark and daylight had brought its renewal of hope. She went on, doubtful that her navigation could have been accurate during her imprisonment in the clouds. She did some mental calculations and reckoned that she should soon reach land.

After some thirteen hours airborne, Amelia saw the conjunction of clouds and hills, but without an altimeter she could not be sure that she could go high enough through the clouds to clear the hills ahead.

She made landfall in Donegal, though she did not know it, and as she wanted to ascertain her position more certainly, as well as find a safe landing place, she veered north and followed the little railway track she spotted until she reached the misty green hills that came sweeping down to the river mouth west of Londonderry.

Amelia came to earth in a field at Culmore, Derry, after a crossing lasting only thirteen and a half hours. The petrol would have lasted another hour or two, but would the aeroplane? She was not to know the answer to that, yet she felt pretty thankful to be out of the air, much as she loved it.

'I had not the faintest notion where I was,' she said afterwards, describing her final minute after sighting land. 'I circled around your city awhile in the hope of locating an aerodrome, but finding none I decided to make for the pastures and land on the nearest suitable field.

'My flight is finished and my ambition is realized. I have crossed the Atlantic alone, and that is what I set out to do. I did it really for fun, not to set up any records or anything like that.'

In fact the flight broke two records. Apart from becoming the first woman to fly the Atlantic solo, Amelia had set up a new fastest time for the near-2,000-mile journey.

President Hoover telegraphed his own and America's congratulations.

Amelia Earhart remained what she had always been, cultured and modest, through her triumphs in the air and the adulation they brought her.

Amelia did not stop there. She had had an urge to outdo the record long-distance flight by a woman, at that time held by Ruth Nicholls with a cross-continent flight in the United States.

So she took a Lockheed Vega, similar to the one in which

she had crossed the Atlantic three months earlier, and started from Los Angeles on 24 August 1932. Her aim was a non-stop trip right across the changing landscape of the States, with its mountains and wildernesses, to Newark, New Jersey, the nearest convenient point for New York.

She covered over 2,500 miles in slightly over nineteen hours. Together with her transatlantic triumph, her two flights demonstrated more than any others just what the future held for civil aviation. And the fact that a woman had proved it was possible to get from Los Angeles to London in less than two days underlined its bright future.

On 11–12 January 1935, she undertook a flight which had cost many airmen their lives, from Honolulu to Oakland, California. The distance was 2,408 miles. She flew solo in a Lockheed Vega with a Pratt and Whitney Wasp engine, and touched down at Oakland just 18 hours 16 minutes after leaving the Polynesian world of the Hawaiian Islands.

The last and most ambitious flight that Amelia Earhart attempted was to encircle the globe at the equator. She set out with Captain Fred Noonan on this round-the-world challenge. But on 2 July 1937, their aeroplane was lost in the Pacific Ocean between British New Guinea and Howland Island.

Amelia Earhart will never be forgotten.

14

COLONEL BLACKER

looking down on Everest

MANY men had dreamed of looking down on Everest before the first flight over it was achieved in 1933, and many men had lost their lives attempting to conquer it by air.

Throughout the ages, men had stood and stared at the tallest peak in the world, wondering how it could be tamed. Only with the advent of aviation did the dream of seeing its summit from close quarters really begin to become a possibility.

Until the 1930s no aircraft engine existed capable of carrying the necessary load to a height of nearly six miles above sea level. Of course this height had been exceeded by ten thousand feet before 1933, but only with the minimum load and in the best possible weather conditions. Neither of these factors would apply to any projected flight over Everest.

Yet man was bound to try and surmount this, the tallest of all peaks: it followed naturally after the crossing of the Atlantic and Pacific and reaching the Poles. Everest was one of the shrinking number of unconquered goals. Its conquest was inevitable; it was just a matter of time.

The development of the Bristol Pegasus engine was what tipped the scales in favour of success. It was a definite advance on any other aeroplane engine previously produced. This nine-cylinder, air-cooled radial engine was supercharged and so was powerful even at high altitudes. It was also wonderfully light for its power. Now it only needed someone to take up the idea of flying over Everest.

Lieutenant-Colonel L. V. S. Blacker conceived the plan early in 1932, but it took over a year of continuous effort to bring it to fruition. As with so many other pioneering ventures, faultless preparation would go a long way toward success. Many people and bodies assisted in their own way. One of the early decisions to be made was the choice of the likeliest aircraft for the job. Two aeroplanes were selected: the Westland PV-3, subsequently called the Houston-Westland, and the Westland Wallace. These were twin-seater biplanes with high undercarriages. Their broad wings gave good climbing qualities and, most important, they could both be adapted to take the new Pegasus engine.

Despite the fantastically low temperatures expected over Everest, the planners decided quite rightly to plump for open cockpits with windshields. The observer's cockpit in each case, however, was panelled above and fitted with side windows for photography.

Oxygen would be supplied to enable the pilots to breathe as easily as possible in the rarefied atmosphere. As photography would be an important purpose of the flight, too, both survey and ciné cameras were fitted for maximum operational convenience.

In addition to the supply of oxygen in pressurized steel cylinders, complete with face masks and flexible tubing, the fliers were to have the benefit of electrically heated clothing. So much special power gear was needed one way and another that the planes incorporated extra dynamos. The pilot and observer in each of the two aeroplanes would wear goggles against the glare produced by the combination of sun and snow that was the climate around Everest.

When these and all the other fittings had been successfully tested and included, the pilots' cockpits looked like the control panels of an advanced airliner instead of a two-seater biplane.

Apart from gauges indicating air-speed, height, revolutions per minute, pressure, petrol, oil, time, and so on, the pilot had valves and regulators for the oxygen supply, a tail incidence wheel, throttle and mixture control – and the main control column.

Three light Moth aeroplanes were flown out to India to aid in the work of conveying films and other records to Calcutta for processing. The two precious Westlands travelled in a more leisurely manner by sea. Blacker chose Purnea as the base for the attempt on the mountain 150 miles away to the north. At Delhi he met the three other principals of the venture, Lord Clydesdale, Flight-Lieutenant D. F. McIntyre and Air Commodore P. F. M. Fellowes.

They flew from there to Purnea in two of the little Moths, and on the way they rounded a lower mountain to get their first glimpse of Everest. It was like the largest gem in a set of three, the others being Kanchenjunga and Makalu. Bejewelled in brilliant whiteness, they looked radiant yet remote. The thought of flying over them seemed almost blasphemous.

This was only a preliminary survey to see how the airfield and hangars were progressing at Purnea. As things seemed to be going ahead well, they started back to Karachi to collect the two Westland aircraft. At Allahabad, however, a strong storm blew up in the night while they were asleep and smashed the little parked Moth aeroplane to pieces. They completed the trip back rather ignominiously by train and then in a borrowed aeroplane! That was the only incident at this stage, though, and the four men and their two large Westlands were soon safely back at Purnea.

From then on it was a question of making the final test flights, and of waiting for the weather to improve. They let up balloons to trace the course and speed of the winds over the upper slopes, but these were swept up and away at speeds well above the maximum safe level decreed. Often the wind was twice as fast as the 30–40 miles an hour regarded as reasonable in the circumstances. It seemed as if they would be marooned at Purnea for ever, waiting for the wind to drop.

The little Moths made daily survey flights to report on the current conditions, and finally came word that visibility was as clear as it ever would be, though the wind was still too high at 57 miles an hour.

'It's no good waiting any longer,' Clydesdale said. 'Let's have a shot tomorrow.'

The whole outfit had to work like madmen to make the

elaborate last-minute preparations: film, oxygen, fuel, instruments tested, telephones between pilot and observer. Everything had to work. There would be no chance of a successful forced landing where they were going, not near the top anyway.

The two Westlands took off on the morning of 3 April 1933. Piloting the Houston-Westland was Lord Clydesdale, with Blacker acting as observer. Flight-Lieutenant McIntyre flew the Westland Wallace, with an air cameraman of Gaumont-British, Mr S. R. Bonnett, filming the flight.

The accuracy of the comparison between the control panel of the biplanes and that of an airliner can be judged as true when one learns that Blacker had to make no fewer than forty-six checks on take-off.

They flew on and up through the inevitable Himalayan haze that lay over everything between 6,000 and 19,000 feet. They rose out of this like a swimmer suddenly breaking surface. Now the haze was a white carpet below them tempting them to try and land on it.

Kanchenjunga soared up to the right. Then they saw Everest itself, with Makalu on its right. The trio of proud peaks rose sheer from the haze which the clear sky seemed to tint pale purple. Billowing like bunting or white smoke from the crest of Everest was the famous rafale, indicating strong winds over the top. The aviators hardly needed the evidence of this snowstream to tell them of the violent air currents around them.

They were aware that no one had seen the glaciers below. On the two biplanes bumped, with Everest their goal. Suddenly the first aircraft was drawn down in a down-draught. Clydesdale and Blacker fell 2,000 feet in what seemed the click of a camera shutter, as if the gods were angry at these mere mortals' daring.

The Houston-Westland approached the gaunt crags of the South Col of Everest. They had watched those weather balloons buffeted about over the glaciers like lost souls and now they were in the same situation. Could the powerful

Pegasus engine resist the surging airstream? There was a brief battle between the elements and the engine. The Pegasus gradually got the upper hand and they regained the lost height, but the incident had shaken them.

Soon they were sweeping around the ragged arc that traced the summit. They were literally on top of the world. So close did the Westland sneak to the summit that Blacker stopped taking pictures to wonder whether the tail skid would in fact foul the mountain. Then he went on with his photography.

They were up there all right, but the next problem would be to get down. Clydesdale headed the aeroplane into the face of the wildcat wind as it streamed snow off the summit. The aeroplane had a nominal speed of 120 miles an hour. The speed of the wind must have been as much because for a moment stalemate reigned.

Then the aeroplane ran the whole gauntlet of the plume – that trail torn from the top of the world. Ice crackled into the open cockpit so savagely that it cracked the windows.

They had seen enough of the roof of the world for one day, so after a rapid circuit or two of the peak, they aimed back to Purnea, landing shortly before half past eleven, after three and a half hours of the most unusual flying ever known to man. There had been one or two snags with the machines but nothing that could not be corrected.

They set their hearts on a second shot at Everest, although they had conquered it. Meanwhile the immediate idea was for a flight over the next most majestic of the Himalayas, Kanchenjunga. This was to test the vertical cameras again, as the Everest snowdust haze had marred some plates taken the first time. This secondary trip would also allow them to perfect the rest of the complicated gear of all kinds.

They made the flight the following day, the aeroplanes coming clear of the haze exactly as before at around 19,000 feet. Now only the gaunt shapes of the Himalayan giants broke the infinity of sky and space.

On Kanchenjunga, as throughout the range, they saw the strange contrast between the snow-stacked side and the darker, rocky side where the wind had blown everything off the steep slopes.

When they got to about 28,000 feet, the summit was shrouded in cloud, rather an unnerving outlook. They flew on and then the leading aircraft suffered a series of sudden shocks. It tilted, twisted, shuddered and spun till they felt they must dive downward, but then the pilot regained control and all was well.

They hesitated and then decided not to try and cross the summit, so they veered off for Purnea, rather like a fighter breaking off a dogfight.

They got away from the mountain but then the unexpected occurred, as it so often did in those early days of flying. They had to make two forced landings. The second one was at Dinajpur, where within ten minutes a throng of thousands of locals had encircled the plane, which had run out of petrol. Most of the people had never seen an aeroplane before that moment.

When the photographs of both flights had been processed, they found that the 19,000-foot high haze had in fact blotted out most of the vertical shots, so confirming the need for another flight over Everest. The aviators did not really need an excuse for it, but were glad they could definitely justify the trip.

McIntyre meanwhile had thought of a good way of utilizing the weather. This involved a different approach to the mountain, by taking advantage of the favourable east wind up to 18,000 or 19,000 feet. Then they could swing up to the summit.

This second flight over Everest took place on 19 April. The estimate of the wind force at 24,000 feet was 88 miles an hour, and up to 100 miles an hour around the summit. In other words, nearly the speed of the aeroplane. Despite this disquieting report, they took off that morning.

A hundred miles from Makalu they took a clear infra-red picture of the peak, though at that stage the summit of Everest still lurked behind a bank of cumulus clouds. They felt in a world apart, here with their heads in the clouds. It was a dizzy, dramatic experience.

Blacker concentrated on his vertical photographs for almost an hour. This brought them immediately in front of Everest, some twenty miles ahead.

Then some strange things happened. The electric plug in Clydesdale's oxygen heater suddenly started vibrating. He grappled with the pins to try and tighten the plug, and had no sooner got it right again than a clamping screw attached to the survey camera began to be affected by the intense cold.

One or two other items all gave signs of stress at that increasing altitude and decreasing temperature. Just as long as nothing vital breaks, thought Blacker. He was taking oblique-angle shots of some of the unexplored ridges in the range south-west from Everest. This was exciting stuff. They soared towards Everest, now twelve miles distant.

The long raked rafale lay ahead, visible in a clear deep blue sky through the framework of struts and wires of their little mechanical world. The rafale blew off the summit like the smoke from a funnel of some celestial liner. Away to their right, behind the rear wing strut, Makalu looked almost as high as Everest.

Now they were only two miles away. The crags of the South Col looked like irregular pyramids.

They were over the summit almost before they realized it. Blacker went on juggling with the three cameras in his control: the vertical survey camera, the oblique still camera for pictures from the side, and the ciné camera. He used up all his plates as the aeroplane pounded around the crest quite happily. The whole string of connected peaks in the huge Himalayan range reminded them of a host of rocky islands strung in a cloudy sea.

This blanket base of cloud really insulated the fliers from all external influences. Their world at that moment was the uninhabited, unknown, unspoilt, Himalayas, but conquered at last – at least from the air.

On they went over Makalu, gasping with wonder as the Khumbu glacier and Arun gorge unfolded below.

The other aircraft took a slightly different course because

the cloud blanket blotted out their environs almost to the foot of Everest. But from both aeroplanes they got good vertical survey pictures, overlapping to form a valuable series of this totally unmapped region. Here was a forerunner of the future: the limitless scope of surveying from the air. Some of these photos revealed a strange patch on the ice-surfaced face of Everest. Like so much else that they had flown over, no one had ever seen this before.

15

WILEY POST

around the world in eight days

AROUND the world in eighty days – that might have been good enough for Jules Verne but Wiley Post had other ideas. He flew around the world in *eight* days: or to be more precise, 8 days, 15 hours and 51 minutes.

Wiley Post and Harold Gatty did this 15,474-mile marathon in a Lockheed Vega with a Pratt and Whitney Wasp engine. The aeroplane was christened *Winnie Mae of Oklahoma* and they followed this route: America, England, Germany, Russia, Alaska, Canada, America.

Wiley Post was an amazing man. He had only one eye and came of Indian blood. After a spell as a parachutist, he graduated to flying, becoming the pilot of an American oil tycoon named F. C. Hall. It was after Mr Hall's daughter that the plane was in fact named.

So these were the ingredients for the flight: a stocky, one-eyed American with a small moustache, Wiley Post; a handsome Tasmanian navigator with special knowledge of many of the air and land conditions along the route, Harold Gatty; and of course the faithful *Winnie Mae* monoplane, emblazoned with the words 'Pathé News, Winnie Mae Round the World Flight.'

The plane had a 450-horsepower engine capable of a maximum speed of 180 miles an hour and it could fly for 3,000 miles without refuelling. This high-wing cantilever plane could normally carry six people and was 27 feet 7½ inches long with a wing span of 41 feet. Wiley Post's seat was up in the nose of the aeroplane, while Gatty had to be content with a very cramped spot that gave him just one or two feet in which to shuffle or stretch. Not ideal for flying around the world.

The two of them underwent more protracted training for this flight than any fliers so far in the history of long-distance records. Among the things Post practised were keeping alert for long periods without falling asleep, and also sleeping at order in short cat-naps. They made their equipment as reliable as possible and included an artificial horizon, blind-flying gear and a special instrument indicating amount of bank and turn.

Eventually they reckoned they were ready. At the dawn of 23 June 1931, 4.55 am New York summer time, they posed for photographers in the misty morning air and then left Roosevelt Field behind them on the first leg of this ambitious attempt.

Their first goal was Harbour Grace, Newfoundland. The *Winnie Mae* ran into midsummer clouds much of the way, but six and three-quarter hours after leaving the magic metropolitan skyline, Gatty told Post, 'Ought to be over Woonsocket in a minute or two.'

The clouds parted as if on cue and the pilot saw the ground, an aerodrome, and a roof bearing the name 'Woonsocket'. Not bad for 1931. So they were poised at Harbour Grace ready for the next little obstacle, the Atlantic Ocean. After a three and three-quarter hour halt for fuel and food, the two round-the-worlders winged effortlessly out over the ocean with a pleasant wind behind them. The plane soon got up to a cruising speed of 170 miles an hour which it maintained with an accuracy equal to Gatty's impeccable course-keeping.

But before they could get too excited about their progress, the weather closed on them and the monoplane met moist vaporous air. Post ascended to 1,800 feet hoping for a break but found none. This was a good initial blooding for their blind-flying equipment, which combined with Gatty's navigation to preserve a remarkable freedom from drift. The navigator was rewarded with a glimpse of the sun as it went down behind them, confirming the course.

For the three hours or so of deepest night, they were once more flying entirely on instruments. With the dawn came the rain. A dreary morning of warm summer rain dulled their senses slightly, though Gatty managed to remain aware of exactly where they were the whole time.

Soon after 11 am Greenwich time, they both noticed a small slit in the rain clouds, like a little ragged hole in a length of grey fabric. Post took the aeroplane through the thick stratum of ' blinding whiteness ' and suddenly saw a ' rugged coast-line '. It was North Wales, slightly over a quarter of a mile below them.

The next thing they viewed was an RAF aerodrome at Sealand, outside Chester. The time was 11.42 GMT. When they landed Post and Gatty both found they were suffering from temporary deafness, brought on by the incessant noise of the engine, but that was their only complaint. While the RAF filled up their tanks, the two men also replenished their stomachs. They took off again within an hour and a half.

On across England, the Norfolk Broads and the North Sea, to Hanover, Germany. This stop was extra to their schedule, and they only stayed while the tanks were topped up to take them on to Berlin. From the moment they touched down at Templehof airport, they were besieged by well-wishers and got fewer hours' sleep than they had hoped.

Early next day they started on the thousand-mile stage to Moscow. The weather forecast was good, but the actual weather they encountered was not and they had a nasty time. At 2,000 feet Post ran into bulging, bursting clouds, so swung down to half that height. The sky still seemed leaden with clouds as they crossed into Poland, East Prussia and finally Russia.

The rain really lashed down and clouds of steam rose off the scalding exhaust pipes. When Post dived low in an effort to out-distance the storm, and to enable Gatty to check their position, he found that more than once the *Winnie Mae* had to dodge the tops of tall chimneys in industrial sectors.

The minarets of Moscow gave them the first inkling of the Orient: this capital that was the half-way house between the west and far east. They had been flying for nearly three days now and were quite tired, but Post and Gatty had no option but to become the guests of the Ossoaviakhim – the Society for Aviation and Chemical Defence. By the time the two men had consumed vodka and heard toasts drunk to their continued success, only two or three hours were left for the sleep they so badly needed.

A strange error delayed their departure from Moscow. The Russians filled the tanks with the requisite number of gallons of petrol, but they went by the Imperial gallon instead of the smaller American one. The difference between the two measures, multiplied by the gallons loaded, made the *Winnie Mae* too heavy. Eventually the Russians drained off the excess fuel and the aviators ascended on one of the trickiest stages of the whole flight: right across to Novo Sibirsk in Siberia.

Mid-way between Moscow and this next stop, the Ural Mountains cut their course at right angles. There was no evading them. Things went swimmingly and the smooth sound of the engine lulled them. The altimeter reported 4,500 feet. The colossal crumpled peaks of the Urals rolled away below, melting into the clouds to the far north.

'We'd be for it if anything happened now,' Post joked.

He had hardly finished speaking when the Wasp engine gave a gulp as if expostulating. The choking protest stammered on. Post reacted promptly and switched the fuel flow to a fresh source. The engine gave a last protest, followed by a purr. It was being fed again. The panic was past.

Losing height roughly parallel to the eastern sides of the Urals they passed Chelyabinsk, the rail junction between Moscow and Siberia, and actually used the legendary Trans-Siberian Railway as a guide. Several hours over the twin rails brought them to Omsk, which they reached at an average speed for the day so far of 176 miles an hour – pretty good going. They had already been in the air for nine hours, but kept on for the remaining two and half hours to bring them to Novo Sibirsk well ahead of their scheduled time. The landing ground was so cut off, however, that they just had to sit and wait for the reception party to drive up by car more than an hour later!

From Novo Sibirsk eastward, the fliers utilized the Trans-Siberian line again as long as visibility allowed. The rail route passed through a kaleidoscopic landscape of hills, forest and wilderness, which would have been dangerous to them in the event of an accident.

They lost track of the railway below sprays of rain, and Post had to carry out his regular manoeuvre of angling up to 9,000 feet to try and top the clouds. He couldn't, so resigned himself to instrument flying back at the familiar 1,500 feet. After yet another cloudburst, the air seemed to clear a little and Post came down low to tag along above the railway to Irkutsk. By now they found it so straightforward to navigate that Gatty could sink down his allotted couple of feet and doze for a time. They clocked up 6 hours 5 minutes from Novo Sibirsk to Irkutsk.

Another little problem of communication greeted them at Irkutsk.They could not find anyone who understood English – let alone American! Finally a sixteen-year-old girl stepped out of the motley crowd of Asiatic onlookers and managed to act as their interpreter. Food, petrol and oil – that was all they wanted. She saw that they got it.

Post and Gatty had calculated in advance that Irkutsk represented the half-way mark, as near as made no odds. They were still doing well and *Winnie Mae of Oklahoma* was holding up marvellously. Irkutsk could hardly have been further from home. But this proved to be the quiet half of the global trip. From then on things began to get awkward.

Whether they could achieve the ten-day time limit they had set remained to be seen.

They set out again after only a couple of hours on the ground. Having flown for so long over land it was strange to find nothing but water beneath them as they cruised comfortably over the great Lake Baikal. It took them an hour to negotiate the lake during which they actually lost all sight of land for a few minutes at the centre.

Their goal was Blagovestchensk. They traversed the remote Yablonoi Mountains and flew on for two or three hundred miles over the stark steppes devoid of any sign of life. It was like being on some strange planet.

By the time they had spanned the Amur river, which was to guide them to their destination, daylight had died. Flares flickered around the aerodrome but the locality had suffered severe rains and the landing ground was so waterlogged that the lights transformed the surface into a single shining pool. Post could have been excused for thinking he had overshot by a few hundred miles and was looking down on a Pacific bay.

He nosed down towards the sodden ground.

The aeroplane had been flying at its usual brisk 170 miles an hour. Now Post had slowed it to 80 miles an hour. It touched the mixture of mud and water. The mud clung in the space between the wheels and the special protective spats, while the water sprayed and splashed as the wheels squelched to a halt. As the aeroplane lost speed and stopped, one wheel sank into a morass of mud. This was not the mud of a ploughed field, for when Post and Gatty got out, they immediately went knee-deep in the mire. They tried to move, but looked like two characters in a slow-motion film. Much more serious, though, was the plight of the aeroplane. Just at that moment it seemed highly unlikely that it could be salvaged from the mud in time for them to beat their self-imposed time-limit.

While Post and Gatty got away, attempts were started to haul the aeroplane out of its predicament. Unless something was done quickly it would sink further. Men floundered frantically about the waterlogged aerodrome, eventually linking a car to the aeroplane to try and tow it out. The driver put the vehicle in gear and let out the clutch. The spinning wheels simply hurled up a dirty spray. Post and Gatty restarted the engines of the aeroplane, but instead of getting out of the mess, it merely sank further into the ooze.

While they awaited a promised tractor, the fliers fell asleep from sheer exhaustion, so the time was not really wasted.

When they awoke the surface seemed to be drying out a bit. As the tractor had not turned up, they tried again to tow it out with the means they had – manual labour plus a couple of horses. The combined strength of all concerned at last managed to drag the aeroplane free of the suction of the mud and on to a firm strip of soil.

The clock had made a complete revolution between their landing and departure from Blagovestchensk, but they were not discouraged, it might have been immeasurably worse. The next hop was to Khabarovask, 363 miles further east, which represented another vital milestone: the vaulting point for the vast 2,000-mile step across the Bering Sea to Alaska and the American continent. They knew this to be the key part of the rest of the journey. If they did it, the worst should be over. They reached Khabarovask easily and prepared for the next leg.

Wiley Post checked everything he could before taking off. He was not too happy about the sparking plugs and so changed them all.

They had another wide sea stretch to bridge before the Bering. This was the Sea of Okhotsk. They flew over it at a height of precisely seventy-five feet to cut down the power of the appalling winds.

They ran into rain, gales and hail as *Winnie Mae* skimmed north-east. The hail peppered against the fuselage, like bullets from some offensive fighter-plane, and then bounced off and plopped into the froth-filled sea.

Post had to drop even lower than seventy-five feet, yet still the gales threatened to engulf them. The aeroplane touched the wave-tops once. The only thing to do was to go right up into the clouds and fly blind, and this he did.

There they got the full fury of the rain, so they rose still higher to 6,000 feet. Later on Gatty said, 'Should be over Kamchatka any minute now.'

They soon were, spotting savage-looking mountains not indicated on any maps. The peninsula of Kamchatka, separating the two seas, was the only land between them and Alaska.

Post did not have to look twice at it to see that they really had no choice but to keep right on for the north-west tip of America.

They sped on over the snow-hatted heads of mountains, until they ran into blue sky and blinding sun. They followed the Russian coast as far as they could, and then branched off across the Bering Sea, cutting the actual water crossing to a minimum. Then they crossed the Date Line; flew into fog; saw St Lawrence Island in the middle of the Bering Sea; and finally grounded at Alaska.

Then another accident happened.

As they taxied along the runway, the wheels started to sink in sand. Post throttled to get lift, but the machine tilted on to its nose. The propeller threshed into the sand and bent the tips of both its blades. The two men soon borrowed a hammer and straightened the bends, but when they had refitted the repaired propeller to the aeroplane and Gatty was swinging it, the engine suddenly spat in a vicious back-fire. The propeller hit him hard on the shoulder, knocked him out for a minute, and bruised his head. Gatty waved aside any assistance, however, and got aboard at once. They picked up a fresh airscrew when they landed at Fairbanks.

They flew another 1,300 miles to Edmonton, Canada, across the Canadian Rockies. Post and Gatty were pretty exhausted by now after a week of concentrated flying.

After that ordeal came another. They came in to land at Edmonton on a flooded flying-ground. Post had to half-fly, half-taxi *Winnie Mae* over to the huddle of hangars, where welcoming crowds surged to meet them.

Neither of the two fliers fancied trying to take off from that aerodrome, reminiscent as it was of the waterlogged Russian field.

'How about using the highway?' suggested a bright official.

The wide main road ran straight for two miles or more. The only snag was to keep it clear, but the police saw to that. Post and Gatty cruised along the Canadian highway at some sixty miles an hour, to the cheers of the local residents. *Winnie Mae* calmly rose clear of the concrete and was gone. Like everything else, she took this in her stride.

And so they pressed on, over Alberta, Saskatchewan, Manitoba, the Great Lakes, Michigan, Pontiac, and Detroit to Cleveland, where they refuelled. Finally, there ahead was the famous New York skyline. They had left it when they headed east and were returning to it from the west.

Aeroplanes paraded on all sides. Wiley Post and Harold Gatty had done it, in 8 days 15 hours 51 minutes.

There is a small postscript to this astounding story. On June 15–22, 1933, Wiley Post did it all over again—*alone*! To girdle the globe solo was impossible, friends told him. He smiled at them and set off. He is quoted as taking with him in the same aeroplane as before a quart of tomato juice, a quart of water, three packets of chewing gum, and a packet of rusk biscuits.

Post flew straight from New York to Berlin this time, the first man ever to achieve this, and then on to Moscow and across Russia. Without a navigator it was all many times more difficult and tiring, but he survived somehow. Once he slumped asleep at the control column. Finally he triumphed over the treachery of fog to cross the Bering Sea and strike Alaska on the home stretch.

The world heard he was overdue at Fairbanks. America was aghast.

Then they learned that he had landed at another airport and damaged his propeller. One was rushed up from Fairbanks by aeroplane and Post fitted it to *Winnie Mae* for the last long voyage across the American continent.

He reached New York in the startlingly short time of 7 days 18 hours 49½ minutes, nearly a day shorter than his previous record with Gatty.

On 15 August 1935, Wiley Post and the comedian Will Rogers were flying a seaplane on a survey trip over Alaska, when fog forced them to land on a river near Point Barrow. As they were taking off, the engine of the seaplane suddenly failed, the machine plunged into the river bank, and both Wiley Post and Will Rogers were killed.

16

SCOTT AND BLACK

the Melbourne Air Race

MILDENHALL to Melbourne was the fantastic flight conceived for the first world-wide air race. This England–Australia contest was made possible by the prizes offered by Sir MacPherson Robertson. The prizes were £10,000, £1,500 and £500 for first, second and third places. The money was to be paid in Australian currency so would be worth about three-quarters of its sterling value. Nevertheless, £7,500 was big enough bait, apart from the prestige attaching to victory.

The total distance was 11,333 miles and to win the race crews would have to be prepared to fly at night for as much of the way as they could. The route to be taken was Mildenhall, Baghdad, Allahabad, Singapore, Darwin, Melbourne.

Stated simply like that, it did not sound as formidable as it really was. From Baghdad onward, for instance, the obstacles became increasingly severe. The white-fringed Persian mountains topped 10,000 feet, while in the Kandahar area before Allahabad, another range rose to 7,000 feet. There was also to be a handicap race over a slightly different course, but the prime interest lay in the speed race itself.

Twenty machines from a number of countries lined up for the start. Apart from British aeroplanes, there were American, Dutch and other nationalities represented, so the event promised to be a truly international epic.

Who would win it? That was the big question at Mildenhall around dawn on Saturday, 20 October 1934.

Despite the early hour, thousands of people from counties far beyond Suffolk fringed the airfield and actually threatened the start altogether. Despite difficulties, the departure moment arrived and the exodus began. The starter's Union Jack dropped as a signal for the first aeroplane to roll forward.

Jim and Amy Mollison were in it, needless to say, and to them went this honour of leading the longest flight of aeroplanes ever assembled for a race. Towards the orange-tinged clouds the Mollisons' De Havilland Comet rose firmly. Its twin Gipsy Six engines bore the machine aloft with its 260 gallons of fuel, sufficient to take it across Europe to the first stopping place. *Black Magic* was the name the Mollisons had chosen for their Comet, one of three similar planes entered.

The second of this trio of Comets was piloted by Cathcart-Jones and Waller. The excitement started there and then as it careered a couple of hundred yards over the Mildenhall grass and then suddenly swung round in the direction of the hangars, and some spectators – as if loath to leave at all.

Flecks of flame from the port engine identified the source of the trouble. The crowd could not help wondering how wise it was for the fliers to go ahead and take off again only two minutes after such a moment.

But with so much going on, they did not have much time to dwell on each individual sortie. A lumbering great American Boeing transport monoplane had risen before the first Cathcart-Jones attempt, piloted by Roscoe Turner and C. Pangborn. On its body it bore the name *Warner Bros*, and on its tail the racing number five. This was bound to be among the possible winners, as the Americans usually did things thoroughly on occasions of this nature.

Stack and Turner in their Airspeed Viceroy taxied along the runway at their allotted forty-five-second interval, and then stopped, turned, and took off nearly quarter of an hour later. The reason: they wanted to collect films of the start of the race and felt this justified the time lost.

Another serious contender among the giants took the air next. It was the large Douglas DC2 flown by Parmentier and Moll.

So the succession of aircraft headed out from East Anglia towards the Low Countries or France. One of the lesser known machines and entries was the Granville Gee Bee monoplane piloted by Wesley Smith and Jacqueline Cochran. The plane took off as if pregnant with petrol! But it got away, that was the main thing.

Davies and Hill in a Fairey IIIF checked out, but a Fairey Fox with Baines and Gilman did not. It seemed that the engine would not start; not a good sign at the outset of 11,000-odd miles. It did get away a little later, after it had missed its turn.

Meanwhile all the others of the twenty had risen safely, including the third of the DH Comets with C. W. A. Scott and Thomas Campbell Black. Like the others, their machine was powered by a pair of Gipsy Six engines. Both these men had long aeronautical experience, Scott having beaten the record between England and Australia three times in 1931–2, while Black had piloted from London to Nairobi thirteen times. In 1931 he had rescued a German flier stranded on the unlikely locale of an island in the Upper Nile. They were two strong favourites for the race.

These were the twenty machines and their fliers:

Airspeed Courier, Stodart and Stodart
Airspeed Viceroy, Stack and Turner
Boeing Transport, Roscoe Turner and Pangborn
Desoutter, Hansen and Jensen
DH Comet, Cathcart-Jones and Waller
DH Comet, Jim and Amy Mollison
DH Comet, Scott and Black
DH Dragon, Hewett and Kay
Douglas DC2, Parmentier and Moll
Fairey Fox, Baines and Gilman
Fairey Fox, Parer and Hemsworth
Fairey III, Davies and Hill
Granville Gee Bee, Wesley Smith and Jacqueline Cochran
Klemm Eagle, Shaw
Lockheed Vega, Woods and Bennett
Miles Falcon, Brook
Miles Hawk, MacGregor and Walker

Monocoupe, Wright and Polando
Panderjager, Geysendorfer and Asjes
Puss Moth, Melrose

One of the most fascinating facts of the race was that the new 200 mile-an-hour Douglas DC2 flown by Parmentier and Moll was actually carrying three paying passengers and air mail. This American-designed plane was run by Royal Dutch Air Lines and it represented the result of several years' aviation development and a taste of the future to the air-minded onlookers of 1934.

Some planes had radio transmitters; others didn't.

So news was spasmodic, even sparse, considering that twenty machines were in the air and would continue to be all day unless or until something happened to any of them.

The first trouble reported was when the Airspeed Viceroy came down at Abbeville. Stack and Turner did manage to get through to Le Bourget at 3.22 pm, but that kind of time would not put them in the forefront, certainly not of the speed race, anyway. It seemed that fog coupled with internal installation faults had caused the delay.

Another casualty over France was the Fairey Fox with Parer and Hemsworth, a leaking radiator forcing them down in the Boulogne area. They were not expected to resume until the following day, Sunday.

The weather was not kind. Brook in his Miles Falcon landed at Plesses, near Paris, while two other machines also grounded 'somewhere in France': the Fairey III south of the capital through lack of fuel, and the DH Dragon also near Boulogne. Five temporary casualties already out of a total of twenty aeroplanes. In a race like this, it would not be easy to make good any serious setbacks. Someone was bound to get through more or less flawlessly.

What was happening up among the leaders? The Mollisons looked as if they might add to their illustrious careers with a win. Whether or not they could keep it up, they certainly clocked in first at Baghdad, closely followed by another of the DH Comets, flown by Scott and Black. So British aircraft were well to the fore 2,530 miles from Mildenhall. Could they sustain such a pulverizing pace? The answer would be known in another two days. Meanwhile the third Comet had not yet arrived at Baghdad, but was expected fairly soon.

Right up there among the giants, as anticipated, was the Douglas DC2 airliner, which touched down at 11.10 pm GMT. Parmentier and Moll in the Douglas DC2 had averaged about 170 miles an hour despite three stops so far.

It was at 11.55 pm, three-quarters of an hour later, that the fourth place was filled by the Panderjager. Spurred on by this arrival, Parmentier and Moll took the Douglas DC2 up again at precisely midnight GMT.

The big Boeing turned up next, refuelled in half an hour, and pressed on for Karachi. The only snag that Roscoe Turner and Pangborn seemed to have experienced was a failure to contact Baghdad by their radio, but they could hardly grumble at that. A technical note: the engine cowlings bore not a single speck of oil.

The third Comet did eventually arrive at the airport an hour after the Middle East dawn had given shape and substance to the alien skyline of Baghdad's buildings.

'Are you all right?' came anxious enquiries of Cathcart-Jones and Waller. For it was widely known that since the Comets had speed they also had a restricted flying time – and this last one of them was overdue by fuel consumption.

Like every competitor in the race, they had an intriguing yarn to tell of their adventures to date.

They had been airborne at 17,000 feet for much of the earlier part of the flight, keeping course by their instruments. Across Hungary and Romania, the weather had been bad, and they only made out the ground beneath them an hour before the Comet flew over the Black Sea. That sojourn over the inland sea was just one more awkward stretch on an outward leg full of unfriendly hazards.

Then they overshot Baghdad!

They actually managed to put down safely 100 miles or so

beyond the check point and decided to wait for the first light to head back to find it. They touched down at Baghdad with precisely two gallons of petrol left!

Cathcart-Jones and Waller maintained this record of mishaps by taking off from Baghdad at 5.57 am – and landing there again at 6 am! The starboard engine was registering nil oil pressure. The lubrication fault behind this resulted in a partial engine seizure, which involved changing a cylinder and piston. Six hours elapsed before they finally got away again, into gathering clouds.

As the leaders raced across the border into India, the Mollisons were still ahead of Scott and Black. Jim Mollison was taking the southerly, coastal route of the two to Allahabad, and the husband-and-wife team set up a record run to Karachi.

But for the bulk of the field, things went less smoothly, the casualties continuing remorselessly. Woods and Bennett brought their Lockheed Vega in for a landing at Aleppo, Syria. They had already had trouble at Athens and perhaps as a result of this, the plane crash-landed at Aleppo, turned right over on its back, and they were forced to give up hope of continuing.

Stack and Turner in the Airspeed Viceroy were also out of the real race, though they were still flying the course. Baines and Gilman in the Fairey Fox, and Shaw in the Klemm Eagle, reached Rome.

The Desoutter had checked in at Aleppo, with the DH Dragon and the Puss Moth en route for that point. So the field was stretched out across thousands of miles already.

Everyone was doing their best, but the spotlight swung inevitably to the east, where a change was about to take place in leadership.

The Mollisons stayed only an hour at Karachi, anxious to hang on to their lead. But less than half an hour further on, towards the next intermediate point of Jodhpur, they realized that they could not continue. Their undercarriage had sustained a heavy knock when landing earlier on, and now it refused to retract. The Mollisons discussed the situation and decided to be prudent and head back to Karachi, the nearest airport, for they knew that the Comet's engines were liable to overheat if the wheels could not be brought up.

They managed to carry out a makeshift repair on the undercarriage, but then fog enveloped Karachi and they lost more hours before they were finally able to get away. They felt particularly frustrated after having set up such a startling time from England to India.

Meanwhile the second Comet took advantage of the Mollisons' mishap to drone eastward on the 2,300-mile hop from Baghdad to Allahabad at an average of 190 miles an hour. They clocked this speed despite a detour.

So now Scott and Black took over the lead. The interminable delay for the Mollisons also meant that the next contenders picked up a place.

For Scott and Black the third leg of the five-stage flight would take them to Singapore. The trim Comet took off from Allahabad, bound on the 2,210-mile ordeal to Singapore. They chose the direct route, instead of the overland one via Rangoon and Bangkok. So Scott and Black headed southeast, left India behind, and commenced that long, agonizing span across the Bay of Bengal – graveyard for fliers throughout the century.

It landed as scheduled at Jask, near the mouth of the Persian Gulf; at Karachi; and then Allahabad. Reports said that it had even returned to Allahabad a few minutes after take-off to collect a passenger missing at the time of departure. A combination of service with speed. By the time the Douglas DC2 touched down at Allahabad, Scott and Black were in the vicinity of the Andaman Islands, flying east of them across the gigantic bay. They had a clear lead of a thousand miles or more, but fortunes could change dramatically in a race of this length, like the sudden switch of the aeroplanes occupying third and fourth places. The Dutch Panderjager was now on the air lane to Allahabad, flying in third position. They had left Karachi a long way behind. As they came down to land at Allahabad, something went wrong with the landing gear and the heavy craft actually struck the ground with a wing-tip and an airscrew blade. For a few seconds the machine made frantic vacillations and then the pilot expertly brought it to a safe halt. But the undercarriage failure put them out of the running. It was all the more galling as they were then under two hours behind the tail of their compatriots in the Dutch Douglas DC2. But that was the luck of the race. The aeroplane with fewest faults would probably win.

Roscoe Turner and Pangborn were glad of this good luck, but they had some bad fortune to offset it. They left Karachi in the big Boeing, but due to a breakdown in radio contact with Allahabad they found themselves flying off course. Fuel and time were both getting short when they at last located Allahabad and got down to earth almost at the last gasp from the fuelless engine. After a hasty fill-up, the Americans were off after the leaders, glad to have a full tank and to be flying third now.

Spread across the Asiatic skies, the survivors of this gruelling cross-continental race roared on. The Comet was really living up to its name, searing for Singapore in a trail of glory. Could anyone catch it?

Despite thick weather over the Bay of Bengal, they made landfall at Alor Star and headed on for Singapore.

They got there at 10.23 pm GMT on Sunday. They had flown a phenomenal 7,040 miles in under forty hours. They had made the two controlled halts at Baghdad and Allahabad, one unscheduled stop, and here they were at Singapore, still a clear thousand miles ahead of the field. Their overall average was about 176 miles an hour.

Scott and Black stayed only an hour at the British colony, before the fourth and possibly the worst of the five stages in the race – from Singapore to Darwin.

This 2,084-mile hurdle could be flown in a straight line or over more land via Batavia, Rambang and Koepang. They set the Comet direct for Darwin and hoped for the best. At 9.30 am GMT they had covered three-quarters of the distance which left the Timor Sea stretch. After that, of course, they had to fly right across the burning heart of Australia, so the race was far from won. As they aimed out over the sea, they were some eight hours in the lead, by time reckoning.

What was happening behind them? Ill-luck had put the Mollisons out of the 'placed' aeroplanes. They had been troubled by the wind shifts between Karachi and Allahabad and had put down at Jubbulpore, after hobbling along for part of the way on a single engine. They found out there that they had two cracked pistons.

The leaders' only other two possible rivals were chasing them east.

At 7.30 am GMT Parmentier and Moll in the Douglas DC2 were still second as they left Singapore. They passed Batavia, on Java, three hours later at 10.30 am, having made the 570-mile leap at an average of 190 miles an hour.

Now, with Rangoon to the rear and Singapore a long way ahead, the Boeing was still running third behind the Dutch airliner.

Fourth at this moment, as they entered into the third day of the race, was the Comet piloted by Cathcart-Jones and Waller.

So the four planes lay in a rough line somewhere between Rangoon and Darwin.

First, was Scott and Black's Comet; second, the Douglas DC2; third, the Boeing; and fourth, Cathcart-Jones' Comet.

As they progressed, none of those behind the leader could know how near they were to advancing a place. Scott and Black were in trouble over the Timor Sea.

For the whole of that last watery spell, and even earlier, the Comet was flying on one engine. The second engine seized up somewhere before Timor itself. Mile after changeless mile came before their eyes in a hypnotic nothingness of sea and sky. Their remaining Gipsy Six engine held out.

At 11.8 am GMT Scott and Black reached Darwin, having taken 2 days 4½ hours for the trip.

Now they knew they were being followed relentlessly. The Douglas DC2 was between Batavia and Rambang. The Boeing had left Alor Star for Singapore. The leaders lost two and a half hours at Darwin seeing to the engine trouble, before they could go on to Charleville. It would be too cruel to lose now, but they were all too aware that it was no use having reached Australia in record time. The race was still on. It was like a jockey assuming he would win when he was in the lead at Tattenham Corner.

Meanwhile many notable times were being put up by the slower planes. The Miles Hawk had passed Karachi; the DH Dragon was on its way there; the Airspeed Courier was bound for Baghdad. Parer and Hemsworth gave up in their Fairey Fox.

Further back still, the other Fairey Fox took off from Rome but crashed in Apulia and caught fire. Both Baines and Gilman were killed.

Now the news focused on the finish.

At 10.40 pm GMT Scott and Black scorched south-east across the Aboriginal wilds of northern Australia at an average speed of 154 miles an hour to reach Charleville, 1,389 miles from Darwin.

The Douglas DC2 had not closed up and a thousand miles and more still separated them from Scott and Black. In fact, the Boeing was gaining ground on the Dutchmen. Roscoe Turner and Pangborn in the Boeing left Singapore eight hours behind the Douglas DC2, but they were going to try and cut the gap by flying direct to Darwin. Cathcart-Jones and Waller in the second Comet had been placed somewhere between Allahabad and Singapore.

MacGregor and Walker in the Miles Hawk had got to Jodhpur.

But back in Australia, the excitement was passing fever-point. Scott and Black were on the last straight run, the 787 miles from Charleville to Melbourne.

Still flying steadily on one engine for much of the way, they covered those final miles efficiently, effortlessly – or so it seemed.

At 5.30 am GMT, Monday, 23 October 1934, the De

Havilland Comet 34 appeared over the Flemington Racecourse, Melbourne, the official finish of the great race. They made the regulation two circuits, landed at Essenden, were brought back to Flemington, and it was all over – bar the fantastic reception.

So Scott and Black had flown from Mildenhall to Melbourne in less than three days: 2 days, 22 hours and 59 minutes.

After that it was simply a matter of waiting to see who would come second and third. The Douglas DC2 reached Charleville several hours after Scott and Black had won, but then the race took a dramatic turn. After leaving Charleville, they started going off course across the desert hinterland of Australia, and finally they had to make a forced landing at Albury racetrack. The Boeing was already heading for Charleville, but a long way behind.

The Dutch decided to play safe and wait for daylight. Then they faced the climax of their whole flight. They had to take off from the confined space of Albury racecourse, knowing that the race depended on it. Could the great Douglas DC2 do it? They were only an hour or so away from Melbourne – but if they failed they might as well still be at Mildenhall.

They got up safely from the short straight track and reached Melbourne second, with the Boeing coming in third.

What a race!

17

JEAN BATTEN

four famous flights

PRETTY and petite Jean Batten made not one but four famous flights. Perhaps more than any other pioneer, she sums up the spirit of those great, golden days.

Her first success was a flight to Australia, which nearly never took place at all. She had originally set out from Lympne airfield on 22 April 1934, but ran clean out of fuel over Italy. She had to come down in complete darkness near Rome and was lucky to get away without injury. That would have deterred many people, but not Jean.

A fortnight later, on 8 May 1934, this young New Zealander left Brooklands airfield at dawn to attempt to fly solo to Australia. She reached Rome safely this time but only rated a four-line paragraph in the papers to record the event.

Brindisi, Athens, Nicosia, Damascus, Baghdad, Karachi, Allahabad, Calcutta, Rangoon, Singapore, Batavia – these were the bare facts of her progress. Sandstorms around Baghdad had put her back a bit. It was only when she reached Batavia that public interest began to quicken. She arrived there in thirteen days and had six and a half days left to beat Amy Mollison's record for the flight of nineteen and a half days.

Next day she had got to Kuepang. By then her clothes were very grimy as she had had to carry out her own repairs to the Gipsy Moth aeroplane at Kuepang and Rambang.

This was in fact the same sort of aeroplane Amy Mollison used, Jean having the small added advantage of a turn and bank indicator. Apart from that, she had to endure the same extremes of fatigue and concentration. She saw the coast near Darwin after another couple of days from Kuepang, chalking up the record time for a flight by a woman to Australia of fifteen days – four and a half days less than Amy Mollison.

Jean flew on down to Sydney and later visited her home in New Zealand. Then on 8 April 1935, she left Sydney in her little machine, named *Mascot*, to fly back to Britain, hoping to become the first woman to achieve the two-way trip.

Jean wanted to better her own time, of course, and seemed likely to do it for much of the return route, despite heavy headwinds the whole way after crossing the equator.

Her metal Gipsy Moth aeroplane had a Gipsy One engine of 100 horsepower, capable of cruising at a steady 90 miles an hour. In its open cockpit, Jean was exposed to the dust and heat of the tropics and to the cold at all heights over about 4,000 feet. She carried no radio, but navigated with her usual meticulous care across the Timor Sea.

When this veteran Gipsy Moth G-AARB was 250 miles out over the Timor Sea, flying at 6,000 feet, the engine suddenly stopped.

This was what fliers feared most. She was roughly half way across the gap between Australia and Timor, at the point of no return, and it didn't look as if she would be able to go forward or back – only down.

Jean kept her head and did a long, laboured glide down 4,000 feet. Only at 2,000 feet did the engine get going again. Jean did not recall breathing during that dramatic descent.

Tropic thunder and rain were the next items to vary the monotony of the flight. And the day after that, headwinds held her speed down to 60 miles an hour, a dangerously low air-speed.

Over the deserts of India and Iraq the heat was unbearable. Pumping the petrol began to blister her hands badly, and they got progressively worse.

Finally, she reached Littorio airport, Rome, on the afternoon of 25 April, having been forced to land for a short while at Foggia because of engine trouble. This was the same trouble she had suffered at Athens before bridging the Adriatic.

Her time from Darwin was still ahead of her outward flight,

but to better it she had to reach England before the following night of 26 April.

She was unlucky. Jean landed at Marignane, near Marseilles, about noon on 26 April, but was once more delayed, not only through engine trouble this time, but sheer bad luck. One of her tyres was cut by a piece of glass on the runway, putting her back still further. Even the weather took a hand, preventing a flight through the Rhone valley. She realized she could not take off from the Riviera that day, and so resigned herself to failing to outdo her own record of fifteen days.

Foul conditions around Dijon delayed her yet again, and she had to make an unscheduled stop at Abbeville.

Finally she crossed the Channel and jogged to a halt on the same runway which she had left nearly a year earlier. The flight had taken 17 days 15 hours and 15 minutes.

Her hands were a mass of blisters and sores. She was glad it was over.

Jean's next quest was the record for the South Atlantic flight from West Africa to Port Natal, Brazil. Even before she began, she had to fly from England to Dakar. She left England on Armistice Day.

She took off from Port Thies, near Dakar, around dawn on 13 November 1935, in drizzling rain. It was far from tropical weather from that moment on. In fact, so dreary was the dawn that she had to invoke the aid of headlamps from two cars to illuminate the remote West African runway.

Flying a Percival Gull with a 200-horsepower Gipsy Six engine, Jean left the comfort of the coast behind her and headed out westward across the South Atlantic. This stretch of ocean seemed much more isolated than the busier northern route of the Atlantic. She knew she had to do it in a day: 13 November 1935 would be her day of triumph or death.

Flying at times as low as 600 feet, Jean ran slap into appalling air conditions. The Doldrums did not live up to their name that day. The storms seemed to start as soon as she was in them, and never let up for the rest of the long way west. The odd break did occur, but before she could take advantage of it, another one of the endless succession of storms hit her.

The little light aeroplane rattled and battled on. Jean began to wonder whether she had taken on just a bit more than she could cope with this time.

Despite her preoccupation with the weather, she had time to realize that she had never, could never, feel lonelier. She felt more on her own than she believed it was possible for anyone ever to imagine. She was the only human being over this million square miles of sea. If the plane faltered now . . .

Her spirits at the lowest ebb, Jean found herself flying as if blindfolded through one of the major disruptions. The storm was so savage that she fell lower than 600 feet, and then lower still.

Suddenly the only compass she had started to go haywire.

The needle swivelled senselessly through 180 degrees. She felt all was lost without its guidance.

If she could not rely on that, what could she do? She wrestled with the facts. All her mental and physical faculties were being extended to the extreme. It was worse than anything on either of the Australian flights. Almost always when over land there was some hope of a successful forced descent, but here landing would mean certain death.

Jean just had to trust her other instruments and fly on; to try to weather the blows being fisted at her. Finally she came out of the most violent gale and rain, and the compass gradually swung back to normal of its own accord. She breathed a soul-felt sigh of relief. The only explanation she could think of was that some electrical disturbances in the Doldrums must have played havoc with the balance of the compass. Anyway, it was over and she flew on.

Jean Batten landed at Port Natal, Brazil, 13 hours and 15 minutes after taking off through that West African drizzle. She had beaten the previous best solo time by three-and-a-quarter hours.

From there she flew on for Rio, but had not arrived by the evening of 14 November. A ghastly few hours passed before news came of her. Could she be lost after having succeeded in her transatlantic flight? That was the kind of irony that so commonly cut down famous pilots: they survived the worst ordeals to perish in some minor mishap.

A Brazilian military plane was sent to search for her near the city, and the Pan-American Puerto Rican Clipper made a detour to try and trace her. This was wild country.

That morning of 14 November, Jean left Port Natal on schedule but ran right out of fuel when still 175 miles north-east of Rio de Janeiro. Hugging the coast, she came down low near the town of Araruama, and force-landed in a steaming swamp.

She did not fancy staying on this doubtful ground, so tried to taxi to what she thought would be a safer spot. In making this manoeuvre, the airscrew struck a heap of sand and became bent, so that Jean could not take off at all. That evening she telephoned to Rio to report that she was well, and she spent the night at a fisherman's hut. She was picked up next day, still looking smart in her white flying suit, by a military machine.

So her flight set up three new records. The crossing was the fastest for the South Atlantic; she was the first woman ever to fly across it; and her time from Lympne to Port Natal also represented a record.

She was the victim of another minor mishap in a straight-forward flight soon after this. Jean had taken delivery of her aeroplane when it was landed by boat at Southampton from South America and set out to fly to London.

On her flight on 29 December she experienced coil trouble and had to make a hurried forced landing near the South Downs. Aiming for a field at Bepton, near Midhurst, Jean just managed to clear a wood, but struck the top of a low hedge bordering the field.

Fortunately the hedge gave way, but as she came down the undercarriage was damaged. The impact of landing gave her concussion and shock, and she had to receive treatment. Her injuries were not serious, but the incident serves to illustrate how easily an accident can happen.

On 5 October 1936, Jean Batten left Lympne, Kent, on her longest flight – to her home in New Zealand. She was also aiming to beat Broadbent's record for the solo flight from England to Australia.

At Koepang, the final stop before Darwin, she was over a day ahead of this record, allowing an average time for her to cross the Timor Sea. But then a dramatic fault threatened to delay her. There were no spare parts at Koepang, so if an aeroplane sustained damage that could not be repaired, the stoppage could last for days.

As Jean was taking off, a tail tyre suddenly punctured. The split was a serious one, not capable of being patched like an inner tube of a car tyre.

'What on earth can I do?' Jean wondered aloud to the aerodrome officials.

Then one of them had a brainwave, and successfully stuffed a collection of face sponges into the hole!

Jean took off again, complete with the sponges, crossed the Timor Sea, and got to Darwin in the record solo time of 5 days 21 hours and 3 minutes. Broadbent had taken 6 days 21 hours and 19 minutes. So she had beaten the record by a clear day.

But that wasn't the end of it. Despite concern in Australia at her idea of flying the Tasman Sea in a single-engine plane, Jean still signified her intention to go ahead with the original plan to reach New Zealand. With that in mind she headed on for Charleville and Sydney, reaching the latter on 13 October.

The local air authorities still said the Tasman Sea venture was 'over-hazardous'; Jean still said she was going to do it, though she did concede that she should have her engine thoroughly overhauled first. Jean could also expect compass trouble between Australia and New Zealand, due to notable magnetic disturbances. She remembered her recent South Atlantic worries over this question. There always seemed to be problems with the compass over the sea.

Jean took off from Richmond, forty miles out of Sydney, on 16 October, hoping to land in her home country in about eight hours or so. The forecasted westerly winds would help her if they materialized.

She made it. Her actual time for crossing the Tasman Sea was nine and a half hours, and the total time from Richmond to Mangere aerodrome, Auckland, was ten and a half hours. She had beaten the previous best time for the Tasman Sea crossing. Her overall time from England to New Zealand was 11 days 56 minutes. She had achieved the first direct flight from England to New Zealand. It seemed especially appropriate that a New Zealander had done it. Jean was also the first woman to fly the Tasman Sea.

It seemed that Jean had learned a lot from her former misfortunes. Before her long-distance triumphs, Jean was known more by her failures. She had once pancaked precariously into a farm, and on another occasion she had been rescued ingloriously by camels in Baluchistan!

Yet now, in October 1936, she had completed the crossing of the lonely Tasman Sea to bring Britain into direct air contact with her furthermost dominion.

Jean Batten deserved every honour that was accorded her after her four famous flights.

18

BERYL MARKHAM

crash landing in Nova Scotia

STORMS raged above the dark Atlantic as the slim silver and peacock-blue monoplane droned due westward. Suddenly its engine spluttered, coughed – and died. Only 2,000 feet of air separated blonde Beryl Markham from the inky-black ocean below.

She glanced at her watch. It was 10.35 pm. She knew that the cabin petrol tank should have lasted four hours, and that she had taken off only 3 hours and 40 minutes earlier, but the tank was empty.

She had made up her mind to be the first person to fly the Atlantic westward, from England to North America, alone. Only Jim Mollison had done it before – and he had started from Ireland. He had given Beryl his watch before she took off with the warning: 'Don't get it wet!' She hoped she wouldn't.

Apart from Jim's watch, she had a sprig of heather for luck. The only other personal things she carried were three vacuum flasks filled with black coffee, some cold chicken, a bag of fruit and nuts and some notepaper.

Beryl had taken off at 6.55 pm on 4 September 1936 in her Percival Vega Gull aeroplane *Messenger*. Its Gipsy engine would fly it at 150 miles an hour without headwinds. She chose Abingdon aerodrome in Berkshire for her take-off, as it had one of the longest runways in the country.

Watched by a crowd of reporters and photographers, the propeller of the little plane spun. The evening sun flashed on the fuselage as it jolted over the runway. Weighed down with a massive load of petrol, the aeroplane taxied towards the trees fringing the aerodrome.

'She'll never get up,' someone whispered.

But she did, clearing the silhouetted trees by only a few feet. The great volume of fuel was calculated to last a full twenty-four hours. Whether that would be enough to carry Beryl across the Atlantic depended on the strength of the winds she met.

The crossing from east to west was a very different matter from flying the Atlantic the other way – from North America to England – when the pilot normally had tail winds.

Now Beryl was all alone, flying into the sunset. Her cramped cabin would be the limits of her world for the next night and day.

She was aiming for New York, 3,000 miles away. Almost before she had realized it, Beryl had left England, Wales and the Irish Sea behind her.

Broken clouds were scattering drizzle down on Ireland. Through the rainy dusk, she saw the blurred lights of Cork. Then, as the moon dodged behind the thickening cloud, night closed in. She saw Berehaven lighthouse – the very last link with land this side of the Atlantic.

Beryl flew on blind through the wind, the rain and the night, relying on her altimeter and artificial horizon to keep her level at 2,000 feet.

She had no such refinements as radar or radio, yet she was no longer afraid, as she had been during those long days of waiting for the right weather. Not that this was good flying weather. She was heading straight into a 40 mile-an-hour wind which cut down her speed.

At 10.35 pm the engine died. The wings quivered in the wind, and as the engine cut out, the wind was all Beryl could hear. The absence of the comforting drone of the engine jerked her to life.

When an aeroplane stalls, the natural reaction is to pull back the stick to try to climb, but this is wrong. Beryl dipped the nose of the little Gull towards the unseen waterline, and in a few seconds her altitude dropped from 2,000 to 1,000 feet.

She knew she had to reach the cock of the reserve petrol

tank to feed fresh fuel to the engine before it was too late. But before she could turn the cock, she had to find a torch. Beryl groped around in the gloom of the cabin. Her fingers closed around the metal torch and switched it on. She quickly flashed the beam around to locate the cock. The altimeter read 300 feet. Each second it was moving nearer to zero.

At 200 feet she found the cock and turned it.

Only 150 feet above the swirling sea, the petrol flooded through and the motor choked, croaked and exploded into life again, like a man who has just won a battle for breath. Muttering a prayer of thanks, she climbed up into the raging storm again, up to 2,000 feet: she had survived the first crisis.

She flew on, through the miles and the hours of watery darkness. Buffeting headwinds drove the rain straight at the frail-looking cabin. Beryl sat with her hands on the stick, battling to keep the aeroplane level.

With her free hand, she tried to pour out the last flask of coffee into a jug balanced on the floor, but a sudden jerk overturned the jug and the coffee was spilt. That depressed her more than anything else on the flight.

Hour after hour, she watched the needle on the luminous dial of the altimeter flickering like some erratic pulse. The point of no return came and went. Beryl thought of her seven-year-old son, Gerald, asleep in his bed at home.

The storm stayed. Then came dawn, lazing up behind her as the plane droned on westward. Nothing much more happened during the morning. She nibbled some fruit and nuts, all she had left to eat. The tiny cabin allowed her to move her long legs only a matter of inches, and she began to feel twinges of cramp, a hazard she had not bargained for.

At 1 pm on 5 September, an officer on the Holland-America liner, *Spaarndam,* pointed to the sky. 'Look – Mrs Markham,' he said to the man at his side.

They reported by radio that they had seen the speck of her aeroplane about 220 miles east of St Johns, Newfoundland.

She had been flying blind for virtually the whole of her trip. After seventeen hours alone in her cabin she felt frozen and her limbs refused to react promptly. The cold, cramp and petrol fumes were nearly overpowering her. She had to fight off unconsciousness.

Her eyelids started to droop but she jerked herself awake again, shivering. Sleep meant death. Although exhausted and icy cold, she had to go on if she wanted to live.

Ice filmed over the glass of the cabin, adding another hazard.

At 2.04 pm, the Swedish liner *Kungsholm* reported to the world: 'Plane sighted 60 miles east of Cape Race, Newfoundland.'

So the 2,000-mile stretch over the Atlantic was nearly over when she ran into the fog.

That, and the iced-up glass cut Beryl off entirely from the world outside her cabin. She could not distinguish between the fog and the sea as she peered at the misty banks below. At 2.30 pm, she was straining her eyes to see land. She reckoned that the Newfoundland coast must be visible soon – and then she spotted it through a gap in the fog. But a moment later it had gone.

For endless minutes she saw nothing more till she began to wonder if she had imagined those ghostly cliffs etched on the shrouded skyline.

At 2.35 pm they spotted her over St Johns.

She had reached Newfoundland in 19 hours 40 minutes.

St Johns lay below like a ghost-town. She adjusted her course accordingly: first, south to Cape Race, then west to Sydney, on Cape Breton Island.

This was 1936, so Beryl had to work out her course by protractor, compass and map. Somehow she managed to juggle these navigation instruments on her lap, at the same time keeping the stick steady and sparing an eye for the control panel.

Despite the weather and her fatigue, she suddenly felt better. On a southerly course, she passed over Renews, forty-five miles south of St Johns, a quarter of an hour later.

Through the notorious Newfoundland mists she glimpsed the Cape Race lighthouse. An hour after being reported over

St Johns, she was out among the Atlantic rollers again, ten miles off Cape Race.

Beryl Markham had one more long hop to fly, the 300-odd miles across the vast jaws of the St Lawrence to Sydney.

After that, she would have land beneath her nearly all the way and it would be fairly plain flying – or so she hoped. As if to add to her sense of well-being, a following wind arose, and the drone of the engine was soothing.

Suddenly, the engine shuddered.

She was still nearly 100 miles from Cape Breton Island. The faithful Gipsy engine belched black exhaust, ominous against the blue-grey and white sea.

Beryl thought urgently: 'Must be an airlock.' She tried to clear it, frantically turning all the cocks on and off, on and off. They had sharp handles, which cut her hands. Slowly the needle on the altimeter fell as Beryl wrestled with the stick and coaxed the controls to try to keep the Gull airborne.

But the aeroplane was falling fast. In fact, the last petrol tank was nearly empty.

The engine was cutting out and restarting continuously now. It barely managed to support the aeroplane over the water. It struggled on for nearly half an hour, getting more erratic all the time.

Could this be the end? It would be heartbreaking for her to fail now, for suddenly she spotted Cape Breton Island, forty miles ahead.

She reckoned all tanks were about empty now, and she prayed that there was enough fuel in the pipes to take her those last miles.

Time was punctuated by the engine's faltering, knocking notes as, desperately slowly, she drew nearer and nearer to the land.

'I never felt closer to death than when I looked down on the water then,' she said later.

For nerve-racking minutes the monoplane hobbled over the Cape Breton coastline barely keeping itself airborne. Then, like a seabird which has sought and found a place to die, its engine gave a last convulsive cough – and died. Beryl had to land. But all she could see was a wild landscape of black ground broken by huge boulders.

The island swelled up at her as she struggled to twist the aeroplane free of the strange coastline below. Amid the ragged rocks, she saw what looked like a field. She had no choice, the Gull had to ground there. And moments later it crash-landed.

Beryl held her breath as the plane hit. The ' field ' was a boulder-strewn swamp and the Gull's nose ploughed forty feet through the mud.

The propeller snapped off and the port wing sank into the bog. The engine was shattered. The aeroplane juddered to a stop with its tail reared in the air. Beryl hit her forehead on the windscreen and blood gushed down her face.

As her aeroplane began to sink slowly into the bog at Baleine Cove Beryl somehow stumbled clear of the wrecked cabin and collapsed.

An hour later, three fishermen found her waist-deep in mud, holding her face in her hands, just 150 yards from the booming Atlantic breakers.

It was still less than twenty-four hours since she had taken off from Abingdon. She had succeeded in flying the Atlantic alone.

Jim Mollison's watch was still ticking.

'Don't get it wet,' he had told her. She hadn't.

19

SCOTT AND GUTHRIE

the Johannesburg Air Race

THE thrill of an air race is undeniable. One of the most imaginative in the years between the wars was the Portsmouth–Johannesburg £10,000 Schlesinger African Air Race.

Nine aircraft eventually started the race, which was to be handicapped. As it turned out, this levelling device was not necessary to determine the result. The idea was for the aeroplanes all to take off virtually together, or at one-minute intervals, and for the time adjustments to be made when they reached Cairo. The length of the flight was 6,150 miles.

These were the nine entrants:

BA Eagle, Alington and Booth
Double Eagle, Rose
Envoy, Findlay and Waller
Hawk Six, Clouston
Mew Gull, Halse
Mew Gull, Miller
Sparrow Hawk, Smith
Vega Gull, Llewellyn
Vega Gull, Scott and Guthrie

At 6.30 am on 29 September 1936, just three years before the War, the first of the nine machines started to shudder across Portsmouth airport in the clear coastal light. As dawn edged brighter over Langstone Harbour and Hayling Island, the mutter of many engines disturbed the morning air and Findlay and Waller's large Envoy aeroplane led the way into the turquoise sky.

But out of these nine aeroplanes competing for the £10,000

prize, only one actually completed the course to Johannesburg. Aeroplanes were hardly reliable even as recently as 1936.

Once they had faded south-eastward, people could only wait for news from the first check point, Belgrade. The handicaps, incidentally, ranged from the scratch allowance of Halse in his Mew Gull to Alington and Booth's 21 hours 58 minutes 12 seconds. The former plane had a formula speed of 200.86 miles an hour, the latter only 116.93 miles an hour. The rest lay between these extremes, with Miller's Mew Gull rated at almost exactly the same scratch figure as Halse.

The first real news of the nine came when Flight-Lieutenant Tommy Rose had some fuel feed difficulty and had to lose time by landing his Double Eagle at Linz to put it right. The delay did not set him back too much and he was reported at Vienna a little later.

Yugoslavia was as excited as Britain at the prospect of being on the route of the remarkable race. Belgrade airport was humming with animation all morning waiting for the first aeroplane to show up out of the north-west sky. Which would it be? The weather was still fine and clear, though the prospects indicated rain as the day wore on.

By noon the atmosphere had become electric. Then at 12.08 pm a Mew Gull winged overhead. As both of the Mew Gulls in the race were capable of 200 miles an hour it was not surprising that one led the field at this stage. The brilliant red tone of the body identified it as Halse's machine. While the petrol flowed into his tank he gulped down coffee and munched sandwiches, answering questions between bites.

At 12.32 pm, just twenty-four minutes after landing, Halse took off again with a useful lead. He felt especially pleased that Miller had not landed in the other Mew Gull by the time he had left. Already people were beginning to wonder where Miller had got to.

At 12.42 pm Clouston touched down in his Hawk Six, and took off again in the staggeringly short time of twelve minutes. He was not away, though, before the third of the aeroplanes appeared.

'Still not Miller,' people were saying.

It was Findlay and Waller in the large Envoy that had left Portsmouth first. This aeroplane was carrying a crew of four, which gave the crew the chance of streamlining the various duties both in the air and on the ground. Despite this asset, they were set back ten minutes by a minor fault in the starboard engine. At 1.16 pm Waller was satisfied and they were off.

Meanwhile in the lull while waiting for the next arrivals, Belgrade airport heard from Vienna that both Tommy Rose in his Double Eagle and Llewellyn flying one of the two Vega Gulls had come and gone.

Forty-eight minutes ticked on as the spectators scanned an empty sky, now filling with clouds. Only five aeroplanes had been accounted for, and two of these had yet to reach Belgrade. The main worry was for Miller in the other Mew Gull, which would have more or less used all its fuel by this time.

Victor Smith in his Sparrow Hawk landed at 2.04 pm, looking frozen from his hours in an exposed cockpit. He was not even wearing an overcoat or extra flying jacket. Someone lent him a leather jacket at Belgrade, which saved the situation for him.

C. W. A. Scott and Giles Guthrie checked in at 2.08 pm with the first of the Vega Gulls.

By 2.15 pm the weather had broken, with rain soaking into the airfield, spattering off the runway. It was then that they saw Tommy Rose overhead in his Double Eagle. He had arranged to circle the aerodrome to report his presence there and then fly straight on to Athens. He did this, but a minute or two later he reappeared over Belgrade, having decided to get the weather forecast for further down the Balkans.

Just before Scott and Guthrie left, the other Vega Gull landed through the rain to bring Llewellyn in to report at 2.23 pm. It was just a couple of minutes after this, in fact, that not only Scott and Guthrie but Victor Smith also took off for Athens.

Since the Scott and Guthrie team and Llewellyn were both in machines carrying a comfortable handicap, they were well placed. It seemed that the race might develop into a scrap between them. Llewellyn cut his time on the ground to a mere thirteen minutes and so when he wobbled up into a fresh crosswind, he was only eleven minutes behind his rivals. Rose, meanwhile, received his weather forecast, decided he had plenty of petrol, and continued in the Double Eagle.

So now seven aeroplanes had reported and two were still adrift. During the long afternoon at Belgrade, the organizers got word that the aeroplane having the advantage of the biggest handicap, the BA Eagle flown by Alington and Booth, had had to make a forced landing in the wilds of Bavaria near Regensburg. In doing so, the clatter over the broken ground damaged the undercarriage and they were out of the race.

The last of the competitors unaccounted for one way or another was still Miller. Belgrade knew nothing, nor did London. He should have been in Yugoslavia hours earlier than this in his fast Mew Gull. All afternoon they searched the sky, now completely covered in cloud, for the white-painted plane.

Finally at 4.24 pm it calmly came in from the west.

'What happened to you?' came the chorus of questions to Miller as soon as he had got out of the cockpit. They were all relieved but anxious to learn what on earth had delayed a 200-mile-an-hour machine for this time. All the others still in the race were well on their way to Athens, or had already got there.

The answer was simple and sad. Miller had flown to within thirty miles or so of Belgrade airport and then been forced to ground as he had run right out of fuel. As this was Yugoslavia, no telephone existed in the outlying area he had chosen. Miller had been mortified to see three rival aeroplanes passing quite low over his fuelless machine and although he waved feverishly no one spotted him.

The wait was a long one, too, for it took him over four hours to get a dozen gallons of petrol to take him on to the check point. Between noon and 4.15 pm he had been begging the Yugoslavs for petrol and when it appeared it could not be described as high-octane by any stretch of the imagination. Ten minutes later he was at Belgrade airport telling his sorry story. The sadder ending was that Miller withdrew from the race as the Mew Gull seemed to be using much more fuel than he had expected, and he might run out repeatedly.

On they flew past Athens to Almaza airport, Cairo. Here the order was still the same as when the leading three left Belgrade.

Halse brought the scarlet Mew Gull down on African soil at 7.07 pm. Clouston was a fraction over half an hour later at 7.39 pm. Findlay and Waller brought up the third place another half hour behind that, at 8.08 pm.

So a mere hour separated the three leaders after the Portsmouth–Belgrade–Cairo stage of 2,249 miles. Halse in his Mew Gull had spanned this considerable distance in something like twelve and a half hours, averaging around 180 miles an hour. Little wonder the Mew Gull was scratch in the handicap ratings.

Three hours after Findlay and Waller, Scott and Guthrie clocked in just four minutes ahead of Llewellyn. So the Vega Gull rivalry had hotted up to an amazing degree; especially as Llewellyn had a thirteen-minute handicap advantage over his colleagues in their sister machine.

So five had got to Cairo, two were out of it, and two remained to reach Cairo. The hop across the Mediterranean might have meant disaster. Until news came, no one could tell. Tommy Rose had not left his troubles at Linz. His Double Eagle was having a harrowing time and so was he. A succession of air locks in the fuel system threatened all across Europe to finish his part in the race. He had landed at a small string of unscheduled spots since setting out that morning. Already the dawn take-off seemed weeks distant.

Finally Tommy Rose coddled the Double Eagle over the

sea to reach Almaza. He signalled to his co-flier in the other cockpit, Bagshaw by name, and they came down to land at Cairo airport. As the machine taxied across the Egyptian field, one of its undercarriage legs crumpled and collapsed.

Rose brought the aeroplane to a halt all right, but the damage would take time to rectify, and time was what Rose could not spare, so he gave up, never imagining that in fact he might still have had a chance however long it took to repair.

There were six competitors left.

Victor Smith had had to come down at Skopje, between Belgrade and Salonika, with oil tank trouble. Much later he made Cairo safely, but was losing a lot of ground he would be unlikely to recover – especially with a persistent fault.

Meanwhile the five leaders forged south for Khartoum. Still scarcely an hour parted the first three, Halse, Clouston, Findlay and Waller.

The pace was gruelling. Something had to give, and at Khartoum it did. The engine of Clouston's Hawk Six seized up. It was not an insurmountable accident, but the time delay would probably forfeit him any chance of victory.

Halse seemed to be drawing away from the field now. He had got to Khartoum under an hour ahead of Clouston, who flew much the slower aeroplane. So Captain S. S. Halse was in the lead and Clouston had come to grief, or at least his engine had.

Who would take over second and third places now?

It was like some gigantic horse race run over a 6,150-mile course, a Grand National of the air with half the hurdles already covered.

Llewellyn touched down a mere two minutes ahead of Scott and Guthrie. The two Vega Gulls were neck and neck. Attention still naturally focused on the first aeroplane, though, and Halse continued to set a scorching rate southward. It seemed almost a formality that he would win. But as with horse races, the result was always in doubt right until the winning post.

Findlay and Waller ran into trouble 300 miles north of

Khartoum, necessitating a forced landing for fuel in the inhospitable region between Wadi Halfa and Kareima. They reported at Khartoum that night, though, a while after dark.

Halse left Kisumu about 8 am on the second day.

The rivals in their respective Vega Gull aircraft were still slogging it out, only a few miles apart.

By midday on the second day, Halse was still comfortably ahead, having passed Mbeya, in Tanganyika. He seemed certain to win, or at least arrive first.

Meanwhile Victor Smith was still dogged by the bad luck of the oil-tank trouble and he fumed at Salonika, with his engine practically choked with oil. Clouston, too, was still delayed at Khartoum awaiting a spare piston. Victor Smith did eventually get as far as Khartoum, a remarkable achievement in the circumstances, but as far as the race was concerned he was virtually out of it.

The toll was mounting all the time. Next followed one of the shocks of the race. Halse was getting tired by the time he approached Salisbury. Twenty miles short of the town, he started trying to identify the airfield, but smoke from veldt fires was camouflaging the landscape and the air was very bumpy. He decided to come down in a field at Bomboshaw.

The light was ebbing fast now and he wanted to land quickly. As the aeroplane touched down, it tumbled over on to its back. Halse dislocated his shoulder and received other minor knocks. The aeroplane was badly hit, however, and he had to abandon any idea of finishing the flight.

Scott and Guthrie had leaped from fourth to first place in a matter of hours. While Halse was meeting disaster, they had flown from Khartoum to Kisumu. The other Vega Gull, however, had taken the more conventional and less dangerous route to Juba west of their course.

Scott and Guthrie left Kisumu at 1.30 that second afternoon and got to Abercorn some six hours later.

The other Vega Gull refuelled on schedule at Juba, flew over Entebbe at 2.15 pm and was not reported again for a long time. Findlay and Waller were still in the race, with the remarkable Envoy, reaching Entebbe by the end of that day.

On the third day Scott and Guthrie left Abercorn at 3.45 pm for the final long hop down to the Rand Airport, Germiston, Johannesburg. The sun rose as they sped southward. By 8.30 am the population of Bulawayo saw the Vega Gull winging on its way with only 450 miles to go to the finish. The race had more shocks in store yet, though.

Llewellyn and his colleague came down low trying to find Abercorn aerodrome, with their fuel almost all expended. The combination of bad weather and lack of fuel forced them into an emergency – and the second Vega Gull crash-landed.

There were three aeroplanes left in the race: Scott and Guthrie in the other Vega Gull; Findlay and Waller flying the Envoy; and Clouston still at Khartoum.

Clouston did in fact later fit his spare piston and fly on gamely to Entebbe, 2,300 miles from the start. Then he took off with the prospect of bad weather ahead and was overdue for quite a time before they found him some 150 miles south of Salisbury with the wreckage of his Hawk Six. He was still smiling, perhaps simply glad to be alive.

So to the dramatic finale, between the remaining two competitors.

As Scott and Guthrie battled on south beyond Bulawayo, Findlay and Waller had left Entebbe at 5.41 am for Abercorn, Northern Rhodesia. They landed there safely in miserable misty cloud.

The wind veered so that if they wanted to fly straight on they would have to take off up a slope towards a cluster of trees. The Envoy made the run uphill but failed to get enough lift. It caught the tops of the trees and crashed.

Captain Max Findlay and Mr A. H. Morgan, the wireless operator, were killed; Ken Waller and their passenger, Derek Peachey, were slightly injured.

So then there was one.

Scott and Guthrie sailed supremely on, landing at Germiston at 11.34 am. They had done the 6,150-mile flight in 2 days 4 hours 57 minutes. This represented a flying average of 156.3

miles an hour and an overall average of about 116 miles an hour.

So C. W. A. Scott had won not only the Melbourne Air Race, but the Johannesburg one as well. It was a wonderful triumph.